T0201171

Advanced Numerical and Semi-Analytical
Methods for Differential Equations

Advanced Numerical and Semi-Analytical Methods for Differential Equations

Snehashish Chakraverty, Nisha Rani Mahato,
Perumandla Karunakar, and Tharasi Dilleswar Rao

National Institute of Technology
Rourkela, Odisha, India

This edition first published 2019
© 2019 John Wiley & Sons, Inc.

All rights reserved. No part of this publication may be reproduced, stored in a retrieval system, or transmitted, in any form or by any means, electronic, mechanical, photocopying, recording or otherwise, except as permitted by law. Advice on how to obtain permission to reuse material from this title is available at http://www.wiley.com/go/permissions.

The right of Snehashish Chakraverty, Nisha Rani Mahato, Perumandla Karunakar, and Tharasi Dilleswar Rao to be identified as the authors of this work has been asserted in accordance with law.

Registered Office
John Wiley & Sons, Inc., 111 River Street, Hoboken, NJ 07030, USA

Editorial Office
111 River Street, Hoboken, NJ 07030, USA

For details of our global editorial offices, customer services, and more information about Wiley products visit us at www.wiley.com.

Wiley also publishes its books in a variety of electronic formats and by print-on-demand. Some content that appears in standard print versions of this book may not be available in other formats.

Limit of Liability/Disclaimer of Warranty
While the publisher and authors have used their best efforts in preparing this work, they make no representations or warranties with respect to the accuracy or completeness of the contents of this work and specifically disclaim all warranties, including without limitation any implied warranties of merchantability or fitness for a particular purpose. No warranty may be created or extended by sales representatives, written sales materials or promotional statements for this work. The fact that an organization, website, or product is referred to in this work as a citation and/or potential source of further information does not mean that the publisher and authors endorse the information or services the organization, website, or product may provide or recommendations it may make. This work is sold with the understanding that the publisher is not engaged in rendering professional services. The advice and strategies contained herein may not be suitable for your situation. You should consult with a specialist where appropriate. Further, readers should be aware that websites listed in this work may have changed or disappeared between when this work was written and when it is read. Neither the publisher nor authors shall be liable for any loss of profit or any other commercial damages, including but not limited to special, incidental, consequential, or other damages.

Library of Congress Cataloging-in-Publication Data is Applied For
Hardback ISBN: 9781119423423

Cover design: Wiley
Cover image: Courtesy of Snehashish Chakraverty

Set in 10/12pt WarnockPro by SPi Global, Chennai, India

Printed in the United States of America

V10008821_031319

Contents

Acknowledgments *xi*
Preface *xiii*

1   **Basic Numerical Methods**   *1*
1.1   Introduction   *1*
1.2   Ordinary Differential Equation   *2*
1.3   Euler Method   *2*
1.4   Improved Euler Method   *5*
1.5   Runge–Kutta Methods   *7*
1.5.1   Midpoint Method   *7*
1.5.2   Runge–Kutta Fourth Order   *8*
1.6   Multistep Methods   *10*
1.6.1   Adams–Bashforth Method   *10*
1.6.2   Adams–Moulton Method   *10*
1.7   Higher-Order ODE   *13*
    References   *16*

2   **Integral Transforms**   *19*
2.1   Introduction   *19*
2.2   Laplace Transform   *19*
2.2.1   Solution of Differential Equations Using Laplace Transforms   *20*
2.3   Fourier Transform   *25*
2.3.1   Solution of Partial Differential Equations Using Fourier Transforms   *26*
    References   *28*

3   **Weighted Residual Methods**   *31*
3.1   Introduction   *31*
3.2   Collocation Method   *33*
3.3   Subdomain Method   *35*
3.4   Least-square Method   *37*

3.5   Galerkin Method   *39*
3.6   Comparison of WRMs   *40*
        References   *42*

4   Boundary Characteristics Orthogonal Polynomials   *45*
4.1   Introduction   *45*
4.2   Gram–Schmidt Orthogonalization Process   *45*
4.3   Generation of BCOPs   *46*
4.4   Galerkin's Method with BCOPs   *46*
4.5   Rayleigh–Ritz Method with BCOPs   *48*
        References   *51*

5   Finite Difference Method   *53*
5.1   Introduction   *53*
5.2   Finite Difference Schemes   *53*
5.2.1   Finite Difference Schemes for Ordinary Differential Equations   *54*
5.2.1.1 Forward Difference Scheme   *54*
5.2.1.2 Backward Difference Scheme   *55*
5.2.1.3 Central Difference Scheme   *55*
5.2.2   Finite Difference Schemes for Partial Differential Equations   *55*
5.3   Explicit and Implicit Finite Difference Schemes   *55*
5.3.1   Explicit Finite Difference Method   *56*
5.3.2   Implicit Finite Difference Method   *57*
        References   *61*

6   Finite Element Method   *63*
6.1   Introduction   *63*
6.2   Finite Element Procedure   *63*
6.3   Galerkin Finite Element Method   *65*
6.3.1   Ordinary Differential Equation   *65*
6.3.2   Partial Differential Equation   *71*
6.4   Structural Analysis Using FEM   *76*
6.4.1   Static Analysis   *76*
6.4.2   Dynamic Analysis   *78*
        References   *79*

7   Finite Volume Method   *81*
7.1   Introduction   *81*
7.2   Discretization Techniques of FVM   *82*
7.3   General Form of Finite Volume Method   *82*
7.3.1   Solution Process Algorithm   *83*
7.4   One-Dimensional Convection–Diffusion Problem   *84*
7.4.1   Grid Generation   *84*

7.4.2 Solution Procedure of Convection–Diffusion Problem *84*
References *89*

8 Boundary Element Method *91*
8.1 Introduction *91*
8.2 Boundary Representation and Background Theory of BEM *91*
8.2.1 Linear Differential Operator *92*
8.2.2 The Fundamental Solution *93*
8.2.2.1 Heaviside Function *93*
8.2.2.2 Dirac Delta Function *93*
8.2.2.3 Finding the Fundamental Solution *94*
8.2.3 Green's Function *95*
8.2.3.1 Green's Integral Formula *95*
8.3 Derivation of the Boundary Element Method *96*
8.3.1 BEM Algorithm *96*
References *100*

9 Akbari–Ganji's Method *103*
9.1 Introduction *103*
9.2 Nonlinear Ordinary Differential Equations *104*
9.2.1 Preliminaries *104*
9.2.2 AGM Approach *104*
9.3 Numerical Examples *105*
9.3.1 Unforced Nonlinear Differential Equations *105*
9.3.2 Forced Nonlinear Differential Equation *107*
References *109*

10 Exp-Function Method *111*
10.1 Introduction *111*
10.2 Basics of Exp-Function Method *111*
10.3 Numerical Examples *112*
References *117*

11 Adomian Decomposition Method *119*
11.1 Introduction *119*
11.2 ADM for ODEs *119*
11.3 Solving System of ODEs by ADM *123*
11.4 ADM for Solving Partial Differential Equations *125*
11.5 ADM for System of PDEs *127*
References *130*

12 Homotopy Perturbation Method *131*
12.1 Introduction *131*

12.2   Basic Idea of HPM   *131*
12.3   Numerical Examples   *133*
        References   *138*

**13   Variational Iteration Method   *141*
13.1   Introduction   *141*
13.2   VIM Procedure   *141*
13.3   Numerical Examples   *142*
        References   *146*

**14   Homotopy Analysis Method   *149*
14.1   Introduction   *149*
14.2   HAM Procedure   *149*
14.3   Numerical Examples   *151*
        References   *156*

**15   Differential Quadrature Method   *157*
15.1   Introduction   *157*
15.2   DQM Procedure   *157*
15.3   Numerical Examples   *159*
        References   *165*

**16   Wavelet Method   *167*
16.1   Introduction   *167*
16.2   Haar Wavelet   *168*
16.3   Wavelet–Collocation Method   *170*
        References   *175*

**17   Hybrid Methods   *177*
17.1   Introduction   *177*
17.2   Homotopy Perturbation Transform Method   *177*
17.3   Laplace Adomian Decomposition Method   *182*
        References   *186*

**18   Preliminaries of Fractal Differential Equations   *189*
18.1   Introduction to Fractal   *189*
18.1.1 Triadic Koch Curve   *190*
18.1.2 Sierpinski Gasket   *190*
18.2   Fractal Differential Equations   *191*
18.2.1 Heat Equation   *192*
18.2.2 Wave Equation   *194*
        References   *194*

19   Differential Equations with Interval Uncertainty *197*
19.1   Introduction *197*
19.2   Interval Differential Equations *197*
19.2.1   Interval Arithmetic *198*
19.3   Generalized Hukuhara Differentiability of IDEs *198*
19.3.1   Modeling IDEs by Hukuhara Differentiability *199*
19.3.1.1 Solving by Integral Form *199*
19.3.1.2 Solving by Differential Form *199*
19.4   Analytical Methods for IDEs *201*
19.4.1   General form of nth-order IDEs *202*
19.4.2   Method Based on Addition and Subtraction of Intervals *202*
          References *206*

20   Differential Equations with Fuzzy Uncertainty *209*
20.1   Introduction *209*
20.2   Solving Fuzzy Linear System of Differential Equations *209*
20.2.1 α-Cut of TFN *209*
20.2.2   Fuzzy Linear System of Differential Equations (FLSDEs) *210*
20.2.3   Solution Procedure for FLSDE *211*
          References *215*

21   Interval Finite Element Method *217*
21.1   Introduction *217*
21.1.1   Preliminaries *218*
21.1.1.1 Proper and Improper Interval *218*
21.1.1.2 Interval System of Linear Equations *218*
21.1.1.3 Generalized Interval Eigenvalue Problem *219*
21.2   Interval Galerkin FEM *219*
21.3   Structural Analysis Using IFEM *223*
21.3.1   Static Analysis *223*
21.3.2   Dynamic Analysis *225*
          References *227*

Index *231*

Acknowledgments

The first author greatly appreciates the patience, support, and encouragement provided by his family members, in particular, his wife Shewli, and daughters Shreyati and Susprihaa. The book may not have been possible without the blessings of his parents late Sh. Birendra K. Chakraborty and Smt. Parul Chakraborty. The second author's warmest gratitude goes to her family members for their continuous motivation and support, especially Sh. Devendra Mahato, Smt. Premshila, Tanuja, Devasish, and Satish. Further, the third author would like to thank for the support and encouragement provided by all his family members, in particular, his parents Sh. Veeraiah Perumandla and Smt. Alivela Perumandla, and his wife Madhavi as well as sons Charan Sai and Harshavardhan. Finally, the fourth author would like to acknowledge the blessings and motivation provided by his family members, especially his parents Sh. Tharasi Rama Rao and Smt. Tharasi Mahalaxmi. Also second, third, and fourth authors appreciate the inspiration of the first author and his family.

Our sincere acknowledgment goes to the reviewers for their fruitful suggestions and appreciations in the book proposal. Further, all the authors do appreciate the support and help of the whole team of Wiley. Finally, we are greatly indebted to the authors/researchers mentioned in the bibliography sections given at the end of each chapter.

S. Chakraverty
N. R. Mahato
P. Karunakar
T. D. Rao

Preface

Differential equations form the backbone of various physical systems occurring in a wide range of science and engineering disciplines viz. physics, chemistry, biology, economics, structural mechanics, control theory, circuit analysis, biomechanics, etc. Generally, these physical systems are modeled either using ordinary or partial differential equations (ODEs or PDEs). In order to know the behavior of the system, we need to investigate the solutions of the governing differential equations. The exact solution of differential equations may be obtained using well-known classical methods. Generally, the physical systems occurring in nature comprise of complex phenomena for which computation of exact results may be quite challenging. In such cases, numerical or semi-analytical methods may be preferred. In this regard, there exist a variety of standard books related to solution of ODEs and PDEs. But, the existing books are sometimes either method or subject specific. Few existing books deal with basic numerical methods for solving the ODEs and/or PDEs whereas some other books may be found related with semi-analytical methods only. But, as per the authors' knowledge, books covering the basic concepts of the numerical as well as semi-analytical methods to solve various types of ODEs and PDEs in a systematic manner are scarce. Another challenge is that of handling uncertainty when introduced in the model. Moreover, some books include complex example problems which may not be convincing to the readers for ease of understanding. As such, the authors came to the realization of need for a book that contains traditional as well as recent numerical and semi-analytic methods with simple example problems along with idea of uncertainty handling in models with uncertain parameters. With respect to student-friendly, straightforward, and easy understanding of the methods, this book may definitely be a benchmark for the teaching/research courses for students, teachers, and industry. The present book consists of 21 chapters giving basic knowledge of various recent and challenging methods. The best part of the book is that it discusses various methods for solving linear as well as nonlinear ODEs, PDEs, and sometimes system of ODEs/PDEs along with solved example problems for better understanding. Before we address some details of the book, the authors assume that the readers have prerequisite knowledge of calculus, basic differential equations, and linear algebra.

As such, the book starts with Chapter 1 containing preliminaries of differential equations and recapitulation of basic numerical techniques viz. Euler, improved Euler, Runge–Kutta, and multistep methods for solving ODEs subject to initial conditions. Chapter 2 deals with the exact solution approach for ODEs and PDEs. In this chapter, we address two widely used integral transform methods viz. Laplace and Fourier transform methods for solving ODEs and PDEs. Another powerful approximation technique, weighted residual method (WRM), is addressed in Chapter 3 for finding solution of differential equations subject to boundary conditions referred to as boundary value problems (BVPs). In this regard, this chapter is organized such that various WRMs viz. collocation, subdomain, least-square, and Galerkin methods are applied for solving BVPs. A new challenging technique viz. using boundary characteristic orthogonal polynomials (BCOPs) in well-known methods like Rayleigh–Ritz, Galerkin, collocation, etc. has also been introduced in Chapter 4.

Due to complexity in various engineering fields viz. structural mechanics, biomechanics, and electromagnetic field problems, the WRMs over the entire domain discussed in Chapter 3 may yield better results when considered over discretized domain. In this regard, various types of finite difference schemes for ODEs and PDEs, and application of the finite difference method (FDM) to practical problems by using schemes like explicit and implicit have been presented in Chapter 5. Finite element method (FEM) serves as another powerful numerical discretization approach that converts differential equations into algebraic equations. The FDM discussed in Chapter 5 generally considers the node spacing such that the entire domain is partitioned in terms of squares or rectangles, but the FEM overcomes this drawback by spacing the nodes such that the entire domain is partitioned using any shape in general. As such, Chapter 6 is mainly devoted to the FEM and especially Galerkin FEM. Effectiveness of the FEM is further studied for static and dynamic analysis of one-dimensional structural systems. Chapter 7 gives an idea of widely used numerical technique named finite volume method (FVM). Accordingly, brief background, physical theory, and algorithm for solving particular practical problem are addressed in this chapter. A brief introduction to another numerical discretization method known as boundary element method (BEM) is addressed in Chapter 8 along with BEM algorithm and procedure to find fundamental solution.

Some problems are nonlinear in nature resulting in governing nonlinear differential equations. Recently, research studies have been done for solving nonlinear differential equations efficiently and modeling of such differential equations analytically is rather more difficult compared to solving linear differential equations discussed in Chapters 1–8. So, this book may also be considered as a platform consisting of various methods that may be used for solving different linear as well as nonlinear ODEs and PDEs. Though the computation of exact solutions for nonlinear differential equations may be cumbersome, a new class of obtaining analytical solutions, that is semi-analytic approach, has emerged. Generally, semi-analytic techniques comprise of power series or closed-form solutions which have been discussed in subsequent chapters. In this regard, Akbari–Ganji's method (AGM) has been considered as a powerful algebraic (semi-analytic) approach in Chapter 9 for solving ODEs. In the AGM, initially a solution function consisting of unknown constant coefficients is assumed satisfying the differential equation subject to initial conditions. Then, the unknown coefficients are computed using algebraic equations obtained with respect to function derivatives and initial conditions. Further, the procedure of exp-function method and its application to nonlinear PDEs have been illustrated in Chapter 10. Semi-analytical techniques based on perturbation parameters also exist and have wide applicability. As such, Chapter 11 addresses Adomian decomposition method (ADM) for solving linear as well as nonlinear ODEs, PDEs, and system of ODEs, PDEs. In this regard, another well-known semi-analytical technique that does not require a small parameter assumption (for solving linear as well as nonlinear ODEs/PDEs) is Homotopy Perturbation Method (HPM). The HPM is easy to use for handling various types of differential equations in general. As such, a detailed procedure of the HPM is explained and applied to linear and nonlinear problems in Chapter 12. Further, Chapter 13 deals with a semi-analytical method viz. variational iteration method (VIM) for finding the approximate series solution of linear and nonlinear ODEs/PDEs. Then, Chapter 14 confers homotopy analysis method (HAM), which is based on coupling of the traditional perturbation method and homotopy in topology. Generally, the HAM involves a control parameter that controls the convergent region and rate of convergence of solution. It may be worth mentioning that the methods viz. ADM, HPM, VIM, and HAM discussed in Chapters 11, 12, 13, and 14, respectively, not only yield approximate series solution (which converges to exact solution) but they may produce exact solution also depending upon the considered problem.

Emerging areas of research related to solution of differential equations based on differential quadrature and wavelet approach have been considered in Chapters 15 and 16, respectively. Chapter 15 contributes an effective numerical method called differential quadrature method (DQM) that approximates the solution of the PDEs by functional values at certain discretized points. In this analysis, shifted Legendre polynomials have been used for computation of weighted coefficients. Further, in order to have an overview of handling ODEs using Haar wavelets, a preliminary procedure based on Haar wavelet–collocation method has been discussed in Chapter 16. Other advanced methods viz. hybrid methods that combine more than one method are discussed in Chapter 17. Two such methods viz. homotopy perturbation transform method (HPTM) and Laplace Adomian decomposition method (LADM) which are getting more attention of researchers are demonstrated to make the readers familiar with these methods. Differential equations over fractal domain are often referred to as fractal differential equations. Recently, fractal analysis has become a subject of great interest in various science and engineering applications. Often, the differential equations over fractal domains are referred to as fractal differential equations. Accordingly, in Chapter 18, only a basic idea of fractals and notion of fractal differential have been incorporated.

Another challenging concept of this book is also to introduce a new scenario in which uncertainty has been included to handle uncertain environment. In actual practice, the variables or coefficients in differential equations exhibit uncertainty due to measurement, observation, or truncation errors. Such uncertainties may be modeled through probabilistic approach, interval analysis, and fuzzy set theory. But, probabilistic methods are not able to deliver reliable results without sufficient experimental data. Therefore, in recent years, interval analysis and fuzzy set theory have emerged as powerful tools for uncertainty modeling. In this regard, Chapter 19 deals with the modeling of interval differential equations (IDEs). Interval analysis modeling of IDEs by Hukuhara differentiability, analytical methods for IDEs along with example problems are addressed in this chapter. A simple technique to handle fuzzy linear differential equations with initial conditions taken as triangular fuzzy numbers is studied in Chapter 20. In fuzzy set theory, a fuzzy number is approximately represented in terms of closed intervals using the α-cut approach. As such, interval uncertainty is sufficient to understand since it forms a subset of fuzzy set. In this regard, FEM discussed in Chapter 6 has been extended for differential equations having interval uncertainties in the last chapter viz. Chapter 21, where we focus on solving uncertain (in terms of closed intervals) differential equations using Galerkin FEM viz. interval Galerkin FEM. Finally, static and dynamic analyses of uncertain structural systems have also been discussed in this chapter.

In order to emphasize the importance of chapters mentioned above, simple differential equations and test problems have been incorporated as examples for easy understanding of the methods. Few unsolved problems have also been included at the end for self-validation of the topics. For quick and better referencing, corresponding bibliographies are given at the end of each chapter. We do hope that, this book will prove to be an essential text for students, researchers, teachers, and industry to have first-hand knowledge for learning various solution methods of linear and nonlinear ODEs and PDEs. Moreover, one can easily understand why and how to use uncertainty concept in differential equations when less or insufficient data are available. As such, this book brings a common platform for most of the newly proposed techniques for solving differential equations under one head along with uncertain differential equations.

2019   *Snehashish Chakraverty, Nisha Rani Mahato,*
Rourkela   *Perumandla Karunakar, and Tharasi Dilleswar Rao*

1

Basic Numerical Methods

1.1 Introduction

Differential equations form the backbone of various science and engineering problems viz. structural mechanics, image processing, control theory, stationary analysis of circuits, etc. Generally, engineering problems are modeled in terms of mathematical functions or using relationships between the function and its derivatives. For instance, in structural mechanics the governing equation of motion

$$m\frac{d^2x}{dt^2} + c\frac{dx}{dt} + kx = f(t) \tag{1.1}$$

associated with Figure 1.1 is expressed in the form of differential equation with respect to the rate of change in time.

Here m, c, and k are mass, damping, stiffness parameters, respectively, and $f(t)$ is the external force applied on the mechanical system.

There exist various techniques for solving simple differential equations analytically. Modeling of differential equations to compute exact solutions may be found in Refs. [1–3]. But, due to complexity of problems in nature, the existing differential equations are rather cumbersome or complex (nonlinear) in nature. Generally, computation of exact or analytical solutions is quite difficult and in such cases, numerical methods [4–8] and semi-analytical methods [9] are proved to be better. In this regard, the numerical and semi-analytic techniques comprising series or closed-form solutions will be discussed in subsequent chapters. Readers interested with respect to accuracy and stability for various numerical methods may study Higham [10]. With the advancement in numerical computing, various software and programming techniques have also been developed that provide efficient platform for solving differential equations numerically viz. MATLAB, Maple, Mathematica, etc.

This chapter presents the recapitulation of basic numerical techniques for solving ordinary differential equations. Few unsolved problems have also been included at the end for self-validation of the topics. However, before we start

Advanced Numerical and Semi-Analytical Methods for Differential Equations, First Edition.
Snehashish Chakraverty, Nisha Rani Mahato, Perumandla Karunakar, and Tharasi Dilleswar Rao.
© 2019 John Wiley & Sons, Inc. Published 2019 by John Wiley & Sons, Inc.

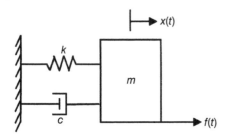

Figure 1.1 Mechanical system.

with numerical methods, we briefly recapitulate the concepts of differential equations.

A *differential equation* is a mathematical relation formulated among function, variables over which it is dependent, and its derivatives. If a function depends on a single variable, then such a differential equation is referred to as ordinary differential equation (ODE), while a function depending on multiple variables result in occurrence of partial differential equation (PDE).

1.2 Ordinary Differential Equation

The general form of an nth order ODE is given by

$$G^n(t, x) = G\left(t, x, \frac{dx}{dt}, \ldots, \frac{d^{n-1}x}{dt^{n-1}}, \frac{d^n x}{dt^n}\right) = 0 \tag{1.2}$$

subject to initial or boundary conditions. The differential equations subject to initial conditions are generally referred to as initial value problems whereas differential equations subject to boundary conditions over a domain Ω where $t \in [t_0, T]$ forms boundary value problems. A detailed discussion on existence and uniqueness of the solution of ODE may be found in Refs. [1–3].

If $G^n(t, x)$ is a linear function of x and its derivatives, then Eq. (1.2) is considered as a linear ODE, otherwise nonlinear. Generally, a nonlinear differential equation consists of product of the dependent variable x with itself or its derivatives. This chapter is evenly divided by focusing on solving ordinary differential equations using Euler, improved Euler, Runge–Kutta, and multistep methods [5, 7]. Laplace and Fourier transform methods for solving differential equations are included in Chapter 2 and numerical solutions of ODE and PDE using difference schemes over grids have been discussed in Chapter 5.

1.3 Euler Method

Let us consider a first-order ODE,

$$\frac{dx}{dt} = F(t, x) \tag{1.3}$$

subject to initial condition $x(t_0) = x_0$.

The numerical methods for solving differential equations are generally obtained initially by partitioning the independent variable $t \in [t_0, T]$ into n finite subintervals at steps $t_0, t_1, \ldots, t_{n-1}, t_n = T$ having equal step size $h = \Delta t = t_i - t_{i-1}$ for $i = 1, 2, \ldots, n-1$.

In terms of Taylor's series expansion, the solution function $x(t)$ of Eq. (1.3) is obtained using

$$x(t + \Delta t) = x(t) + \Delta t \frac{dx}{dt} + \frac{\Delta t^2}{2!} \frac{d^2 x}{dt^2} + \frac{\Delta t^3}{3!} \frac{d^3 x}{dt^3} + \cdots$$

$$\Rightarrow x(t + h) = x(t) + h \frac{dx}{dt} + \frac{h^2}{2!} \frac{d^2 x}{dt^2} + \frac{h^3}{3!} \frac{d^3 x}{dt^3} + \cdots \tag{1.4}$$

Equation (1.4) reduces to Euler's equation if we retain the first-order approximation and neglect higher-order terms giving

$$x(t + h) = x(t) + h \frac{dx}{dt}$$

$$\Rightarrow x(t + h) = x(t) + hF(t, x). \tag{1.5}$$

For discrete steps $t_0, t_1, \ldots, t_{n-1}$, Eq. (1.5) may be written as

$$\Rightarrow x_{i+1} = x_i + hF(t_i, x_i) \tag{1.6}$$

for $i = 0, 1, \ldots, n-1$. Here, Eq. (1.6) is known as the Euler method or Euler–Cauchy method. Geometrically, the Euler method at each step t_i results in an approximation of $x(t)|_{t=t_i} \approx x_i$ in terms of tangent to curve at t_i where $i = 0, 1, \ldots, n-1$. The geometrical representation of the Euler method is depicted in Figure 1.2.

The error at each step from Figure 1.2 is obtained using $|x(t_i) - x_i|$ for $i = 0, 1, \ldots, n$. Also, the local truncation error [5, 11] at each step is proportional to h^2 having $O(h^2)$ whereas the global error is of $O(h)$, where "O" stands for order. The algorithm for the Euler method is formulated in Algorithm 1.1.

Figure 1.2 Geometrical representation of the Euler method.

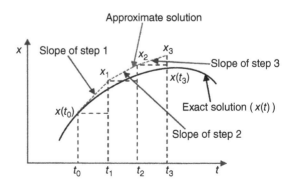

Algorithm 1.1: Euler Method

Input: Differential equation, $F(t, x) = \frac{dx}{dt}$ with initial condition $x(t_0) = x_0$; Domain, $t \in [t_0, T]$; Step size, $h = \frac{T-t_0}{n}$ or Number of subintervals, $n = \frac{T-t_0}{h}$

Step (i): *for* $i = 0$ to $n - 1$

$$t_{i+1} = t_i + h$$

$$x_{i+1} = x_i + hF(t_i, x_i)$$

end *for*

Step (ii): $x(t)|_{t=t_{i+1}} \approx x(t_{i+1}) = x_{i+1}$
Output: Solution $x(t)$

Example 1.1 Solve the differential equation $y' + 2xy = 0$ having initial condition $y(0) = 2$ using the Euler method having step size $h = 0.1$ in the domain $[0, 1]$.

Solution   Here, $\frac{dy}{dx} = y' = -2xy = F(x, y)$, $x_0 = 0$, and $y_0 = 2$. The number of subintervals n is obtained using $n = \frac{1-0}{0.1} = 10$. Using the Euler method given in Eq. (1.6) and corresponding Algorithm 1.1, we get

$$y_{i+1} = y_i + hF(x_i, y_i) \Rightarrow y_{i+1} = y_i - 0.2x_iy_i \tag{1.7}$$

for $i = 0, 1, \ldots, n - 1$. So, the approximate solutions to four decimal places at each step are given in Table 1.1. The comparison of results with exact solutions at x_i for $i = 1, 2, \ldots, 10$ has also been included in Table 1.1.

The exact solution for Example 1.1 may be obtained as $y(x) = 2e^{-x^2}$. Accordingly, the absolute error in Table 1.1 is computed using $| 2e^{-x_i^2} - y_i |$. The corresponding graphical approximation of Example 1.1 comparative to exact solution $y(x) = 2e^{-x^2}$ is depicted in Figure 1.3.

Generally, the Euler method exhibits inaccuracy with an increase in step size h whereas for small h the method provides better accuracy, but it may not be computationally efficient. As such, higher accuracy must be taken into consideration with higher terms in Eq. (1.4),

$$x(t + h) = x(t) + h\frac{dx}{dt} + \frac{h^2}{2}\frac{d^2x}{dt^2} + \frac{h^3}{6}\frac{d^3x}{dt^3} + \cdots$$

Improved Euler and Runge–Kutta methods are discussed in successive sections that exhibit better accuracy compared to the Euler method.

Table 1.1 Approximate solution at x_i using the Euler method.

i	Approximate solution y_i	Exact solution $y(x_i)$	Error $\|y(x_i) - y_i\|$
1	$y_1 = y_0 - 0.2x_0y_0 = 2$	1.9801	0.0199
2	$y_2 = y_1 - 0.2x_1y_1 = 1.96$	1.9216	0.0384
3	$y_3 = y_2 - 0.2x_2y_2 = 1.8816$	1.8279	0.0537
4	$y_4 = y_3 - 0.2x_3y_3 = 1.7687$	1.7043	0.0644
5	$y_5 = y_4 - 0.2x_4y_4 = 1.6272$	1.5576	0.0696
6	$y_6 = y_5 - 0.2x_5y_5 = 1.4645$	1.3954	0.0691
7	$y_7 = y_6 - 0.2x_6y_6 = 1.2887$	1.2253	0.0635
8	$y_8 = y_7 - 0.2x_7y_7 = 1.1083$	1.0546	0.0537
9	$y_9 = y_8 - 0.2x_8y_8 = 0.931$	0.8897	0.0413
10	$y_{10} = y_9 - 0.2x_9y_9 = 0.7634$	0.7358	0.0277

Figure 1.3 Approximate solution of $y' + 2xy = 0$, $y(0) = 2$ using the Euler method and comparison with exact solution.

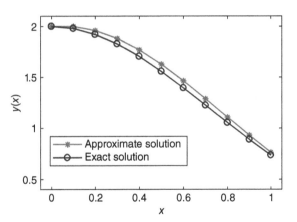

1.4 Improved Euler Method

Let us again consider the first-order differential equation given in Eq. (1.3). The improved Euler method for solution of Eq. (1.3) is a two-step procedure,

$$x_{i+1}^* = x_i + hF(t_i, x_i), \tag{1.8a}$$

$$x_{i+1} = x_i + \frac{h}{2}[F(t_i, x_i) + F(t_{i+1}, x_{i+1}^*)]. \tag{1.8b}$$

In few literature studies, improved Euler method is also termed as modified Euler method or Heun's method. Improved Euler method is also considered as a predictor–corrector method where Eq. (1.8a) acts as the predictor and Eq. (1.8b) acts as the corrector. Eventually, at each step we predict the function value using Eq. (1.8a) and the predicted value is then corrected using Eq. (1.8b). Corresponding algorithm for improved Euler method is formulated in Algorithm 1.2.

Algorithm 1.2: Improved Euler Method

Input: Differential equation, $F(t, x) = \frac{dx}{dt}$ with initial condition, $x(t_0) = x_0$; Domain, $t \in [t_0, T]$; Step size, $h = \frac{T-t_0}{n}$ or Number of subintervals, $n = \frac{T-t_0}{h}$

Step (i): *for* $i = 0$ to $n - 1$

$$t_{i+1} = t_i + h$$

$$x^*_{i+1} = x_i + hF(t_i, x_i)$$

$$x_{i+1} = x_i + \frac{h}{2}[F(t_i, x_i) + F(t_{i+1}, x^*_{i+1})]$$

end *for*

Step (ii): $x(t)|_{t=t_{i+1}} \approx x(t_{i+1}) = x_{i+1}$

Output: Solution $x(t)$

The efficiency of Algorithm 1.2 is verified using Example 1.2.

Example 1.2 Solve a first-order differential equation

$$\frac{dy}{dt} = t^2 - y + 2t \tag{1.9}$$

with initial condition $y(0) = 1$ using improved Euler method for step size $h = 0.2$ in the $[0, 1]$.

Solution The exact solution of Eq. (1.9) is

$$y(t) = e^{-t} + t^2. \tag{1.10}$$

Using Eq. (1.8a), we obtain the prediction

$$y^*_{i+1} = y_i + hF(t_i, y_i) = y_i + 0.2(t_i^2 - y_i + 2t_i) \tag{1.11}$$

where $i = 0, 1, \ldots, n - 1$. Then, the correction is performed using

$$y_{i+1} = y_i + 0.1[t_i^2 - y_i + 2t_i + t_{i+1}^2 - y_i - 0.2(t_{i+1}^2 - y_i + 2t_{i+1}) + 2t_{i+1}].$$

Table 1.2 Approximate solution at t_i using improved Euler method.

	Approximate solution		Exact solution	Error
i	y_i^*	y_i	$y(t_i)$	$|y(t_i) - y_i|$
1	0.8	0.864	0.8587	0.0053
2	0.7792	0.8397	0.8303	0.0094
3	0.8637	0.9213	0.9088	0.0125
4	1.0491	1.1043	1.0893	0.015
5	1.1043	1.3847	1.3679	0.0168

The approximate value to four decimal places at each step is given in Table 1.2. The local error [5, 11] for improved Euler method is of $O(h^3)$ whereas the global error is of $O(h^2)$. Runge–Kutta fourth-order method is discussed in next section which gives better accuracy than the previous two methods.

1.5 Runge–Kutta Methods

The Taylor series expansion in the Euler method is truncated till first derivative whereas the Runge–Kutta method retains truncated terms till second-order derivative. As such, the higher-order term improves the accuracy in Runge–Kutta compared to the Euler method. In this regard, we discuss Runge–Kutta second- (midpoint method) and fourth-order methods in Sections 1.5.1 and 1.5.2, respectively.

1.5.1 Midpoint Method

The explicit midpoint method for the first-order differential equation (1.3) is expressed as

$$x_{i+1} = x_i + hF\left(t_i + \frac{h}{2}, x_i + \frac{h}{2}F(t_i, x_i)\right) \tag{1.12}$$

whereas the implicit midpoint method is expressed as

$$x_{i+1} = x_i + hF\left(t_i + \frac{h}{2}, \frac{1}{2}(x_i + x_{i+1})\right) \tag{1.13}$$

It is worth mentioning that Eq. (1.12) or Eq. (1.13) computes the function at midpoints $t_i + \frac{h}{2}$. As such, the Runge–Kutta second-order method is also referred to as the midpoint method. Also, the explicit second-order Runge–Kutta method is equivalent to improved Euler method.

1.5.2 Runge–Kutta Fourth Order

The fourth-order Runge–Kutta also known as classical Runge–Kutta or RK4 for the differential equation (1.3) is given as

$$x_{i+1} = x_i + \frac{1}{6}(k_1 + 2k_2 + 2k_3 + k_4) \tag{1.14}$$

where

$$k_1 = hF(t_i, x_i),$$

$$k_2 = hF\left(t_i + \frac{h}{2}, x_i + \frac{1}{2}k_1\right),$$

$$k_3 = hF\left(t_i + \frac{h}{2}, x_i + \frac{1}{2}k_2\right)$$

and

$$k_4 = hF(t_i + h, x_i + k_3).$$

Corresponding steps for RK4 are given in Algorithm 1.3.

Algorithm 1.3: Runge–Kutta Fourth-Order Method

Input: Differential equation, $F(t, x) = \frac{dx}{dt}$ with initial condition $x(t_0) = x_0$; Domain, $t \in [t_0, T]$; Step size, $h = \frac{T-t_0}{n}$ or Number of subintervals, $n = \frac{T-t_0}{h}$

Step (i): *for* $i = 0$ to $n - 1$

$$t_{i+1} = t_i + h$$

$$k_1 = hF(t_i, x_i)$$

$$k_2 = hF\left(t_i + \frac{h}{2}, x_i + \frac{1}{2}k_1\right)$$

$$k_3 = hF\left(t_i + \frac{h}{2}, x_i + \frac{1}{2}k_2\right)$$

$$k_4 = hF(t_i + h, x_i + k_3)$$

$$x_{i+1} = x_i + \frac{1}{6}(k_1 + 2k_2 + 2k_3 + k_4)$$

end *for*

Step (ii): $x(t)|_{t=t_{i+1}} \approx x(t_{i+1}) = x_{i+1}$

Output: Solution $x(t)$

Example 1.3 Solve the first-order differential equation

$$\frac{dy}{dt} = y + t\sin(t) \tag{1.15}$$

with initial condition $y(0) = 1$ using the RK4 method for step size $h = 0.2$ within the domain $[0, 1]$.

Solution The exact solution of Eq. (1.15) is

$$y(t) = \frac{1}{2}[3e^t - t\sin(t) - (t+1)\cos(t)] \tag{1.16}$$

Using Algorithm 1.3, we obtain

$$y_{i+1} = t_i + \frac{1}{6}(k_1 + 2k_2 + 2k_3 + k_4)$$

where

$$k_1 = 0.2[y_i + t_i\sin(t_i)],$$

$$k_2 = 0.2\left[x_i + \frac{k_1}{2} + (t_i + 0.1)\sin(t_i + 0.1)\right],$$

$$k_3 = 0.2\left[x_i + \frac{k_2}{2} + (t_i + 0.1)\sin(t_i + 0.1)\right]$$

and

$$k_4 = 0.2[x_i + k_3 + (t_i + h,)\sin(t_i + h)]$$

for $i = 0, 1, \ldots, n - 1$. The approximate value using RK4 rounded to four decimal places is computed and incorporated into Table 1.3.

In case of RK4, the local error is of $O(h^5)$ whereas the global error is of $O(h^4)$ [5, 11]. As such, RK4 is the more efficient numerical method for solving initial value problems compared to Euler and improved Euler methods.

Table 1.3 Approximate solution at t_i using RK4 method.

i	Approximate solution y_i	Exact solution $y(t_i)$	Error $\lvert y(t_i) - y_i \rvert$
1	1.2242	1.2242	0
2	1.5151	1.5151	0
3	1.9035	1.9035	0
4	2.4243	2.4243	0
5	3.1163	3.1164	0.0001

1.6 Multistep Methods

Numerical method is referred to as single-step method when y_{i+1} is computed in a single step. The methods mentioned in Sections 1.3–1.5 are single-step methods whereas a multistep method comprises of two or more successive steps.

1.6.1 Adams–Bashforth Method

The multistep formula for Adams–Bashforth method of fourth order is given by

$$x_{i+1} = x_i + \frac{h}{24}[55F(t_i, x_i) - 59F(t_{i-1}, x_{i-1}) + 37F(t_{i-2}, x_{i-2}) - 9F(t_{i-3}, x_{i-3})].$$
(1.17)

Gerald and Wheatley [5] may be referred for the derivation of Eq. (1.17). Further, the local error for the Adams–Bashforth method is of $O(h^5)$ and the global error is of $O(h^4)$.

1.6.2 Adams–Moulton Method

Adams–Moulton method is a predictor–corrector method where we may predict the value of $x(t)$ for the differential equation (1.3) using Eq. (1.17),

$$x_{i+1}^* = x_i + \frac{h}{24}[55F(t_i, x_i) - 59F(t_{i-1}, x_{i-1}) + 37F(t_{i-2}, x_{i-2}) - 9F(t_{i-3}, x_{i-3})].$$
(1.18)

Then the value of $x^*(t)$ is corrected using the Adams–Moulton method. The multistep formula of the Adams–Moulton method of fourth order may be obtained as

$$x_{i+1} = x_i + \frac{h}{24}[9F(t_{i+1}, x_{i+1}^*) + 19F\left(t_i, x_i\right) - 5F\left(t_{i-1}, x_{i-1}\right) + F\left(t_{i-2}, x_{i-2}\right)]$$
(1.19)

for $i = 4, 5, \ldots, n - 1$.

Algorithm 1.4: Adams–Moulton Fourth-Order Method

Input: Differential equation, $F(t, x) = \frac{dx}{dt}$ with initial condition, $x(t_0) = x_0$; Domain, $t \in [t_0, T]$; Step size, $h = \frac{T - t_0}{n}$ or Number of subintervals, $n = \frac{T - t_0}{h}$

Step (i): Obtain starting values using Algorithm 1.1 or 1.2 or 1.3

t_i where $i = 0, 1, 2, 3$

for $i = 4$ to $n - 1$

Step (ii): Predictor (Adams–Bashforth)

$$t_{i+1} = t_i + h$$

$$x^*_{i+1} = x_i + \frac{h}{24}[55F(t_i, x_i) - 59F\left(t_{i-1}, x_{i-1}\right) + 37F\left(t_{i-2}, x_{i-2}\right) - 9F\left(t_{i-3}, x_{i-3}\right)]$$

Step (iii): Corrector

$$x_{i+1} = x_i + \frac{h}{24}[9F(t_{i+1}, x^*_{i+1}) + 19F(t_i, x_i) - 5F(t_{i-1}, x_{i-1}) + F(t_{i-2}, x_{i-2})]$$

     end *for*

Step (iv): $x(t)|_{t=t_{i+1}} \approx x(t_{i+1}) = x_{i+1}$

Output: Solution $x(t)$

Example 1.4   Solve the initial value problem (IVP).

$$\frac{dy}{dx} + y(x - 2) = 0, \quad y(0) = 2 \tag{1.20}$$

using the Adams–Moulton method for step size $h = 0.1$ in the domain $[0, 2]$ with starting values $y(0.2) = 2.8$, $y(0.4) = 3.808$, and $y(0.6) = 5.0266$.

Solution   The exact solution of Eq. (1.20) is

$$y(t) = 2e^{-\frac{x^2}{2} + 2x}. \tag{1.21}$$

Using Algorithm 1.4, the solution is computed and the values are given in Table 1.4.

Finally, the efficiencies of the above numerical methods are verified using Example 1.5.

Table 1.4 Approximate solution using the Adams–Moulton method.

i	Approximate solution y^*_i	y_i	Exact solution $y(x_i)$	Error $\lvert y(x_i) - y_i \rvert$
4	7.3164	7.0061	7.1933	0.1872
5	10.859	9.1517	8.9634	0.1883
6	9.2127	10.803	10.7311	0.0719
7	12.857	12.439	12.3437	0.0953
8	13.802	13.751	13.6419	0.4073
9	14.581	14.601	14.4855	0.1155
10	14.889	14.897	14.7781	0.1189

Example 1.5   Compute and compare the solution of given IVP

$$y' + y + t = 1, \quad y(0) = 1 \tag{1.22}$$

with respect to Euler, improved Euler, Runge–Kutta fourth order, and Adams–Moulton with step size $h = 0.2$ in the domain $[0, 1]$.

Solution   The exact solution for $y' + y + t = 1$, $y(0) = 1$ is $y(t) = 2 - t - e - t$. Accordingly, Eq. (1.22) is solved using Algorithms 1.1–1.4 and those are compared in Table 1.5.

Approximate solution obtained using Euler, improved Euler, Runge–Kutta fourth order, and Adams–Moulton are compared with the exact values and depicted in Figure 1.4.

Further, the absolute error $|y_i - y(t_i)|$ has been computed from Table 1.5 and the error plot is illustrated in Figure 1.5.

It may clearly be seen from Table 1.5 and Figure 1.5 that Runge–Kutta fourth order and Adams–Moulton provide better solution compared to the exact values.

Table 1.5 Approximate solution at t_i using the Euler, improved Euler, RK4, and Adams-Moulton methods.

i	Euler	Improved Euler	RK4	Adams–Moulton	Exact solution $y(t_i)$
		Approximate solution y_i			
1	1	0.98	0.98127	0.98127	0.98127
2	0.96	0.9276	0.92968	0.92968	0.92968
3	0.888	0.84863	0.85118	0.85118	0.85119
4	0.7904	0.74788	0.75067	0.75068	0.75067
5	0.67232	0.62926	0.63211	0.63213	0.63212

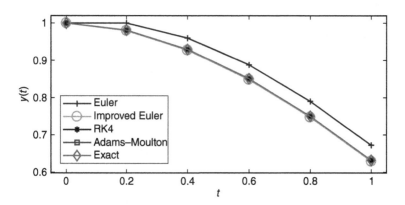

Figure 1.4 Approximate solutions of $y' + y + t = 1, y(0) = 1$.

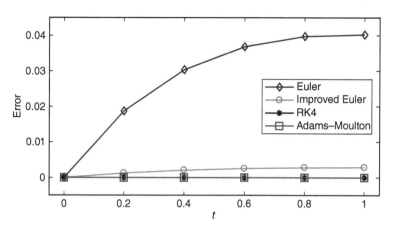

Figure 1.5 Error plot of Example 1.5 using Euler, improved Euler, RK4, and Adams–Moulton methods.

Further, the numerical techniques for first-order ODEs discussed in Sections 1.3–1.6 may be extended to higher order and coupled system of ODEs.

1.7 Higher-Order ODE

Higher-order ODEs may easily be converted to system of first-order ODEs. For instance, consider the nth order ODE given in Eq. (1.2),

$$G^n(t,x) = G\left(t, x, \frac{dx}{dt}, \dots, \frac{d^{n-1}x}{dt^{n-1}}, \frac{d^n x}{dt^n}\right) = 0$$

subject to initial conditions $x(0) = x_0$ and $x^{(i)}(0) = x_0^{(i)}$ for $i = 1, 2, \dots, n-1$. This nth order ODE may be transformed to coupled system of first-order ODEs,

$$\left.\begin{aligned}
\frac{dy_1}{dt} &= F_1(t, y_1, y_2, \dots, y_n) = 0 \\
\frac{dy_2}{dt} &= F_2(t, y_1, y_2, \dots, y_n) = 0 \\
&\vdots \\
\frac{dy_n}{dt} &= F_n(t, y_1, y_2, \dots, y_n) = 0
\end{aligned}\right\} \tag{1.23}$$

by substituting $y_1 = x$, $y_i = \frac{dy_{i-1}}{dt} = \frac{d^{i-1}x}{dt^{i-1}}$ for $i = 2, 3, \dots, n$ and $\frac{dy_n}{dt} = \frac{d^n x}{dt^n}$.

Let us now illustrate the numerical methodology for solving higher-order ODE by converting it to system of first-order ODEs through an example problem.

Example 1.6 Solve second-order IVP

$$x'' - x' + 5x = 4 \tag{1.24}$$

having initial conditions $x(1) = 2$ and $x'(1) = -1$ using Euler and RK4 methods with step size $h = 0.2$ in the domain $[1, 3]$.

Solution Equation (1.24) is reduced to coupled system of first-order IVPs using Eq. (1.23) as

$$\left.\begin{array}{l} \dfrac{du}{dt} = v, u(1) = 2 \\[2mm] \dfrac{dv}{dt} = v - 5u + 4, v(1) = -1 \end{array}\right\} \tag{1.25}$$

where $u = x, v = \frac{dx}{dt}$, and $\frac{dv}{dt} = \frac{d^2x}{dt^2}$.

Above first-order ODEs (1.25) may be solved using methods discussed in Sections 1.3–1.6. In this regard, the Euler method based on Eq. (1.6) for solving system of differential equations transforms to

$$u_{i+1} = u_i + hF_1(t_i, u_i, v_i) \tag{1.26a}$$

and

$$v_{i+1} = v_i + hF_2(t_i, u_i, v_i) \tag{1.26b}$$

where $\frac{du}{dt} = F_1(t, u, v) = v$ and $\frac{dv}{dt} = F_2(t, u, v) = v - 5u + 4$.

Similarly, in case of RK4 the Eq. (1.14) transforms to

$$u_{i+1} = u_i + \frac{1}{6}(k_1 + 2k_2 + 2k_3 + k_4) \tag{1.27a}$$

and

$$v_{i+1} = v_i + \frac{1}{6}(l_1 + 2l_2 + 2l_3 + l_4) \tag{1.27b}$$

where

$$k_1 = hF_1(t_i, u_i, v_i), \quad l_1 = hF_2(t_i, u_i, v_i)$$

$$k_2 = hF_1\left(t_i + \frac{h}{2}, u_i + \frac{1}{2}k_1, v_i + \frac{1}{2}l_1\right), \quad l_2 = hF_2\left(t_i + \frac{h}{2}, u_i + \frac{1}{2}k_1, v_i + \frac{1}{2}l_1\right)$$

$$k_3 = hF_1\left(t_i + \frac{h}{2}, u_i + \frac{1}{2}k_2, v_i + \frac{1}{2}l_2\right), \quad l_3 = hF_2\left(t_i + \frac{h}{2}, u_i + \frac{1}{2}k_2, v_i + \frac{1}{2}l_2\right)$$

$$k_4 = hF_1(t_i + h, u_i + k_3, v_i + l_3) \quad \text{and} \quad l_4 = hF_2(t_i + h, u_i + k_3, v_i + l_3)$$

for $\frac{du}{dt} = F_1(t, u, v)$ and $\frac{dv}{dt} = F_2(t, u, v)$. Using Eqs. (1.26a), (1.26b), (1.27a), and (1.27b), the solutions of differential equation (1.24) are obtained and incorporated into Table 1.6.

The solutions obtained in Table 1.6 using Euler and RK4 are compared with the exact solution $x(t) = e^{\frac{t-1}{2}} \left(\frac{6}{5} \cos\left(\frac{\sqrt{19}t}{2} - \frac{\sqrt{19}}{2} \right) - \frac{16\sqrt{19}}{95} \sin\left(\frac{\sqrt{19}t}{2} - \frac{\sqrt{19}}{2} \right) \right) + \frac{4}{5}$ in Figure 1.6.

Similar to Euler and RK4 techniques, other numerical methods may be extended for higher order and system of first-order ODEs.

Table 1.6 Approximate solution at t_i using Euler and RK4 methods.

		Approximate solution x_i				
		Euler		RK4		
i	t_i	u_i	v_i	u_i	v_i	Exact solution $x(t_i)$
1	1.2	1.8000	−2.4000	1.6595	−2.3931	1.6596
2	1.4	1.3200	−3.8800	1.0567	−3.5733	1.0567
3	1.6	0.5440	−5.1760	0.2647	−4.2366	0.2645
4	1.8	−0.4912	−5.9552	−0.5858	−4.1244	−0.5866
5	2	−1.6822	−5.8550	−1.3227	−3.0896	−1.3242
6	2.2	−2.8532	−4.5438	−1.7605	−1.1535	−1.7626
7	2.4	−3.7620	−1.7993	−1.7377	1.4617	−1.7401
8	2.6	−4.1219	2.4028	−1.1575	4.3368	−1.1594
9	2.8	−3.6413	7.8053	−0.0230	6.9036	−0.0236
10	3	−2.0803	13.8076	1.5416	8.5351	1.5430

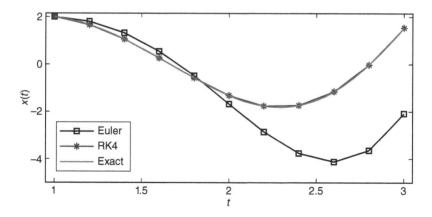

Figure 1.6 Comparison of solutions of Example 1.6 using Euler and RK4 with exact solution.

This chapter is mainly devoted to illustrate the basic ideas of handling initial value ODEs using Euler, improved Euler, RK4, and Adams–Moulton methods. Interested readers may refer the excellent books [4–8] for other numerical methods. Moreover, the methods may be extended to PDEs too.

Exercise

1 Solve the IVPs exactly and compute the error using Euler and improved Euler methods.
   (a) $y' - t^2 - 1 = 0$, $y(0) = 1$, $h = 0.1$, $t \in [0, 1]$
   (b) $y' + 2y^2 - t + 2 = 0$, $y(0) = 0$, $h = 0.25$, $t \in [0, 2]$
   (c) $y' - (y - 2x)^2 - x^3 \sin(x) = 0$, $y(0) = 2$, $h = 0.2$, $x \in [0, 2]$

2 Solve the initial value problem $\frac{dy}{dx} = 3x^3 + 1$, $y(0) = 1$, $h = 0.1$, $t \in [0, 1]$ using RK4 and Adams–Moulton methods.

3 Compute the solution of IVP $y'' + 8xy' + 5xy = 4$, $y(1) = 2$, $y'(1) = -1$ using Euler and RK4 for step size $h = 0.1$ in the domain $x \in [1, 2]$.

4 Evaluate $u'' - u' + u = t$ subject to initial conditions $u(0) = 1$ and $u'(0) = 0$ for step size $h = 0.05$ in the domain $[0, 1]$ using the improved Euler method.

5 Compute the system of ODEs $\frac{du}{dt} = -2u + v^2$ and $\frac{dv}{dt} = -30u + 40v$ subject to initial conditions $u(0) = 0$ and $v(0) = 1$ for step size $h = 0.01$ using Euler, improved Euler, RK4, and Adams–Moulton methods.

References

1 Zwilliger, D. (1998). *Handbook of Differential Equations*, 3e. New York: Academic Press.

2 Hartman, P. (2002). *Ordinary Differential Equations*, 2e. Philadelphia: SIAM.

3 Arnold, V.I. (2006). *Ordinary Differential Equations*, 3e. New York: Springer.

4 Atkinson, K.E. and Han, W. (1985). *Elementary Numerical Analysis*, 3e. New York: Wiley.

5 Gerald, C.F. and Wheatley, P.O. (2004). *Applied Numerical Analysis*, 7e. Boston, MA; Munich: Pearson Education India.

6 Bhat, R.B. and Chakraverty, S. (2004). *Numerical Analysis in Engineering*. Pangbourne: Narosa, Alpha Science International Limited.

7 Hoffman, J.D. (2001). *Numerical Methods for Engineers and Scientists*, 2e. New York: Taylor & Francis.

8 Milne, W.E. and Milne, W.E. (1953). *Numerical Solution of Differential Equations*. New York: Wiley.

9 Ganji, D.D. and Talarposhti, R.A. (2018). *Numerical and Analytical Solutions for Solving Nonlinear Equations in Heat Transfer*, Advances in Mechatronics and Mechanical Engineering Book Series. Hershey, PA: IGI Global.

10 Higham, N.J. (2002). *Accuracy and Stability of Numerical Algorithms*, 2e. Philadelphia: SIAM.

11 Kreyszig, E. (2017). *Advanced Engineering Mathematics*, 10e. New Delhi: Wiley.

2

Integral Transforms

2.1 Introduction

In Chapter 1, we have discussed about basic numerical methods by which one may find the approximate solutions of ordinary differential equations. In this regard, this chapter deals with the exact solutions of ordinary and partial differential equations. The methods we discuss here are integral transform methods. These are used frequently in different fields of engineering and sciences. Especially, Laplace transform (LT) and Fourier transform (FT) are having wide range of applications. Although one may find theories, concepts, and details about these methods in different excellent books [1–8]. But, just to have an idea of these methods, we introduce here the basic concepts of LT and FT. Accordingly, in this chapter, we will address these two methods LT and FT for solving ordinary and partial differential equations.

Integral transform of a function $f(t)$ is defined as below:

$$T(f(t)) = \int_{t_1}^{t_2} K(t,s) f(t)\, dt \tag{2.1}$$

where $K(t, s)$ is called kernel of the transformation.

2.2 Laplace Transform

LT is a well-known integral transform method.

Definition 2.1 Let $f(t)$ be a function of t for $t \geq 0$, then LT of $f(t)$ is defined as

$$F(s) = \mathcal{L}(f(t)) = \int_{0}^{\infty} e^{-st} f(t)\, dt, \tag{2.2}$$

where $\mathcal{L}$ is the Laplacian operator.

Advanced Numerical and Semi-Analytical Methods for Differential Equations, First Edition.
Snehashish Chakraverty, Nisha Rani Mahato, Perumandla Karunakar, and Tharasi Dilleswar Rao.
© 2019 John Wiley & Sons, Inc. Published 2019 by John Wiley & Sons, Inc.

In other way, one may define as below:

If the kernel in Eq. (2.1) is $K(t, s) = e^{-st}$, then the integral transform is called as LT.

Definition 2.2 Let $F(s) = \mathcal{L}(f(t))$ be the LT of $f(t)$, then the inverse LT of $F(s)$ is $\mathcal{L}^{-1}\{F(s)\} = f(t)$, where $\mathcal{L}^{-1}$ is the inverse Laplacian operator.

The list of few formulae and properties of LT that are useful in further discussion are given in Table 2.1 which may easily be obtained using Definition 2.1.

Further details of LT and its applications to various engineering and science problems are available in standard books [1–3, 6].

2.2.1 Solution of Differential Equations Using Laplace Transforms

In this section we study how to use LT for solving ordinary and partial differential equations with the help of three examples. First, we briefly explain how to solve initial value problems using LT [1, 2].

Let us consider a second order linear ordinary differential equation as below:

$$a\frac{d^2y}{dt^2} + b\frac{dy}{dt} + cy = f(t), \tag{2.3}$$

with initial conditions $y(0) = d$, $y'(0) = e$.

Table 2.1 Laplace transforms of $f(t)$.

S. no	$f(t) = \mathcal{L}^{-1}\{F(s)\}$	$F(s) = \mathcal{L}\{f(t)\}$
1	1	$\dfrac{1}{s}$
2	$t^n, \quad n = 1, 2, 3, \ldots$	$\dfrac{n!}{s^{n+1}}$
3	e^{at}	$\dfrac{1}{s-a}$
4	$\cos(at)$	$\dfrac{s}{s^2 + a^2}$
5	$\sin(at)$	$\dfrac{a}{s^2 + a^2}$
6	$e^{at}f(t)$	$F(s-a)$
7	$t^n f(t)$	$(-1)^n \dfrac{d^n F(s)}{ds^n}$
8	Linear property: $\mathcal{L}\{af(t) + bg(t)\} = a\mathcal{L}\{f(t)\} + b\mathcal{L}\{g(t)\}$	
9	LT of nth derivative is $\mathcal{L}\{y^n(t)\} = s^n Y(s) - s^{n-1}y(0) - s^{n-2}y'(0) - \cdots - y^{n-1}(0)$, where $y(0), y'(0), \ldots$ are initial conditions and $Y(s) = \mathcal{L}\{y(t)\}$	

The basic steps involved in solving Eq. (2.3) using LT are discussed below:

Step (i): Take LT on both sides of the differential equation (2.3).

Step (ii): Using differentiation property of LT (Table 2.1) $\mathcal{L}\left\{\frac{dy}{dt}\right\} = sY(s) - y(0)$, $\mathcal{L}\left\{\frac{d^2y}{dt^2}\right\} = s^2Y(s) - sy(0) - y'(0)$, and initial conditions, we obtain $Y(s)$ in terms of s.

Step (iii): Find the inverse LT of $Y(s)$ to get the solution of the initial value problem.

One may solve higher-order differential equations also in the similar fashion. We now consider a simple initial-value problem to understand the solution procedure of the method.

Example 2.1 Solve the initial-value problem

$$\frac{d^2y}{dt^2} - 4\frac{dy}{dt} + 3y = 6t - 8, \tag{2.4}$$

subject to initial conditions $y(0) = 0$, $y'(0) = 0$.

Solution By applying LT on both sides of Eq. (2.4), we get

$$\mathcal{L}\left\{\frac{d^2y}{dt^2}\right\} - 4\mathcal{L}\left\{\frac{dy}{dt}\right\} + 3\mathcal{L}\{y\} = 6\mathcal{L}\{t\} - 8\mathcal{L}\{t\}$$

$$\Rightarrow s^2Y(s) - sy(0) - y'(0) - 4(sY(s) - y(0)) + 3Y(s) = \frac{6}{s^2} - \frac{8}{s}$$

$$\Rightarrow s^2Y(s) - 4sY(s) + 3Y(s) = \frac{6}{s^2} - \frac{8}{s}$$

$$\Rightarrow (s^2 - 4s + 3)Y(s) = \frac{6 - 8s}{s^2}$$

$$\Rightarrow Y(s) = \frac{6 - 8s}{s^2(s^2 - 4s + 3)} = \frac{6 - 8s}{s^2(s - 1)(s - 3)}$$

Using partial fractions we have

$$Y(s) = \frac{2}{s^2} + \frac{1}{(s - 1)} - \frac{1}{(s - 3)}$$

Taking inverse LT on both sides, we get

$$\mathcal{L}^{-1}\{Y\{s\}\} = 2\mathcal{L}^{-1}\left\{\frac{1}{s^2}\right\} + \mathcal{L}^{-1}\left\{\frac{1}{\{s - 1\}}\right\} - \mathcal{L}^{-1}\left\{\frac{1}{\{s - 3\}}\right\}$$

$$y(t) = 2t + e^t - e^{3t}.$$

Next, we consider a linear nonhomogeneous ordinary differential equation where the solution by other analytical methods like variation of parameters, method of undetermined coefficients, etc. is difficult when the nonhomogeneous term, that is right-hand-side function, is discontinuous or piecewise continuous. But, it may be easier to solve such problems using the LT method if LT of these functions exist.

Example 2.2   Let us consider differential equation

$$\frac{d^2y}{dt^2} + 4y = f(t), \quad y(0) = 0, \quad y'(0) = 0, \tag{2.5}$$

where the function $f(t)$ is defined as $f(t) = \begin{cases} 4, & 0 \le t < 3 \\ 0, & t > 3 \end{cases}$.

Solution   The unit-step function is defined and denoted as

$$u_a(t) = \begin{cases} 1, & t > a \\ 0, & t < a \end{cases}$$

Using the unit-step function notation, the function $f(t)$ can be written as $f(t) = 4(1 - u_3(t))$, where $a = 3$. The plot of this function is given in Figure 2.1. LT of the unit-step function is given as [9]

$$\mathcal{L}\{u_a(t)\} = \frac{e^{-as}}{s}$$

The function $f(t)$ in terms of unit-step function is $f(t) = 4(1 - u_3(t)) = 4 - 4u_3(t)$.

The given differential equation (2.5) may be rewritten as

$$\frac{d^2y}{dt^2} + 4y = 4 - 4u_3(t), \quad y(0) = 0, \quad y'(0) = 0 \tag{2.6}$$

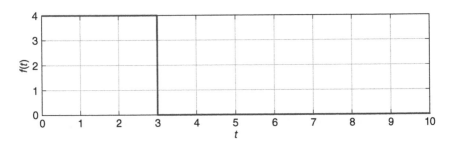

Figure 2.1 Plot of the function $f(t) = 4(1 - u_3(t))$.

By applying LT on both sides of Eq. (2.6), we get

$$\mathcal{L}\left\{\frac{d^2y}{dt^2}\right\} + 4\mathcal{L}\{y\} = 4\mathcal{L}\{1\} - 4\{u_3(t)\}$$

$$\Rightarrow s^2 Y(s) - sy(0) - y'(0) + 4Y(s) = \frac{4}{s} - \frac{4}{s}e^{-3s}$$

$$\Rightarrow Y(s) = \frac{4}{s(s^2 + 4)} - \frac{4e^{-3s}}{s(s^2 + 4)}$$

By applying inverse LT on both sides, we obtain

$$\mathcal{L}^{-1}\{Y(s)\} = \mathcal{L}^{-1}\left\{\frac{4}{s(s^2 + 4)}\right\} - \mathcal{L}^{-1}\left\{\frac{4e^{-3s}}{s(s^2 + 4)}\right\}$$

$$\Rightarrow y(t) = 1 - \cos(2t) - (1 - \cos(2t - 6))u_3(t).$$

In the next example, we apply LT to problem with combination of initial and boundary conditions.

Example 2.3 Consider a nonhomogeneous wave equation as

$$\frac{\partial^2 u}{\partial t^2} = \frac{\partial^2 u}{\partial x^2} + \sin(2\pi x), \quad 0 < x < 1, \quad t > 0 \tag{2.7}$$

subject to the initial and boundary conditions,

$$\left.\begin{array}{l} u(x, 0) = 0, \\ u_t(x, 0) = 0, \\ u(0, t) = 0, \\ u(1, t) = 0 \end{array}\right\} \tag{2.8}$$

Solution   Rewriting Eq. (2.7), we may get

$$\frac{\partial^2 u}{\partial x^2} = \frac{\partial^2 u}{\partial t^2} - \sin(2\pi x) \tag{2.9}$$

By taking LT on both sides of Eq. (2.9), we have

$$\mathcal{L}\left\{\frac{\partial^2 u}{\partial x^2}\right\} = \mathcal{L}\left\{\frac{\partial^2 u}{\partial t^2} - \sin(2\pi x)\right\}$$

$$\mathcal{L}\left\{\frac{\partial^2 u}{\partial x^2}\right\} = \mathcal{L}\left\{\frac{\partial^2 u}{\partial t^2}\right\} - \mathcal{L}\{\sin(2\pi x)\} \tag{2.10}$$

LTs of partial derivatives with respect to time and that of spatial derivative are defined as below [3]:

$$\mathcal{L}\left\{\frac{\partial^2 u}{\partial t^2}(x,t)\right\} = s^2 u(x,s) - su(x,0) - u_t(x,0) \tag{2.11}$$

$$\mathcal{L}\left\{\frac{\partial^2 u}{\partial x^2}(x,t)\right\} = \frac{d^2 u}{dx^2}(x,s) \tag{2.12}$$

Using initial conditions and substituting Eqs. (2.11) and (2.12) in Eq. (2.10), we obtain

$$\frac{d^2 u}{dx^2}(x,s) = s^2 u(x,s) - \frac{\sin(2\pi x)}{s}$$

$$\Rightarrow \frac{d^2 u}{dx^2}(x,s) - s^2 u(x,s) = -\frac{\sin(2\pi x)}{s} \tag{2.13}$$

Equation (2.13) is second-order nonhomogeneous ordinary differential equation with an independent variable x. The solution of the differential equation (2.13) is expressed as

$$u(x,s) = u_h + u_p,$$

where u_h is the solution of the homogenous part of Eq. (2.13) and u_p is the particular integral.

The homogeneous solution may be obtained as $u_h(x,s) = c_1 e^{sx} + c_2 e^{-sx}$, where c_1 and c_2 are arbitrary constants.

Let $u_p(x,s) = a_1 \cos(2\pi x) + a_2 \sin(2\pi x)$ be the particular solution of Eq. (2.13). We find unknowns a_1 and a_2 using method of undetermined coefficients [9] as

$$\frac{d^2 u_p}{dx^2}(x,s) = -4\pi^2 a_1 \cos(2\pi x) - 4\pi^2 a_2 \sin(2\pi x).$$

Substituting $u_p(x,s)$ and $\frac{d^2 u_p}{dx^2}(x,s)$ in Eq. (2.13), we get

$$-4\pi^2 a_1 \cos(\pi x) - 4\pi^2 a_2 \sin(\pi x) - s^2(a_1 \cos(2\pi x) + a_2 \sin(2\pi x)) = -\frac{\sin(2\pi x)}{s}$$

$$(s^2 + 4\pi^2)a_1 \cos(2\pi x) + (s^2 + 4\pi^2)a_2 \sin(2\pi x) = \frac{\sin(2\pi x)}{s}$$

$$\Rightarrow a_1 = 0, \quad \text{and} \quad a_2 = \frac{1}{s(s^2 + 4\pi^2)}$$

Therefore, the particular solution of Eq. (2.13) is $u_p(x,s) = \frac{1}{s(s^2+4\pi^2)}\sin(2\pi x)$. General solution of Eq. (2.13) is then written as

$$u(x,s) = c_1 e^{sx} + c_2 e^{-sx} + \frac{1}{s(s^2 + 4\pi^2)}\sin(2\pi x) \tag{2.14}$$

Next, using boundary conditions in Eq. (2.14), we get

$$u(0, s) = 0 \quad \Rightarrow c_1 + c_2 = 0$$
$$u(1, s) = 0 \quad \Rightarrow c_1 e^s + c_2 e^{-s} = 0$$
$$\Rightarrow c_1 = 0 \quad \text{and} \quad c_2 = 0$$

Plugging the values of c_1 and c_2 in Eq. (2.14), we obtain

$$u(x, s) = \frac{1}{s(s^2 + 4\pi^2)} \sin(2\pi x) \tag{2.15}$$

By applying inverse LT to Eq. (2.15), we get

$$\mathcal{L}^{-1}\{u(x, s)\} = \mathcal{L}^{-1}\left\{ \frac{1}{s(s^2 + 4\pi^2)} \sin(2\pi x) \right\}$$

$$u(x, t) = \mathcal{L}^{-1}\left\{ \frac{1}{s(s^2 + 1)} \right\} \sin(2\pi x)$$

$$= \mathcal{L}^{-1}\left\{ \frac{1}{4\pi^2} \left(\frac{1}{s} - \frac{s}{s^2 + 4\pi^2} \right) \right\} \sin(2\pi x)$$

$$= \frac{1}{4\pi^2} \left(\mathcal{L}^{-1}\left\{ \frac{1}{s} \right\} - \mathcal{L}^{-1}\left\{ \frac{s}{s^2 + 4\pi^2} \right\} \right) \sin(2\pi x)$$

$$= \frac{1}{4\pi^2}(1 - \cos(2\pi t)) \sin(2\pi x).$$

Thus, the solution of the given nonhomogeneous wave equation (2.7) using the LT method is

$$u(x, t) = \frac{1}{4\pi^2}(1 - \cos(2\pi t)) \sin(2\pi x).$$

2.3 Fourier Transform

FT is another well-known integral transform method like LT for solving differential equations.

Definition 2.3 The FT of a function $f(t)$ is defined as

$$F(f(t)) = \hat{f}(t) = \int_{-\infty}^{\infty} e^{ist} f(t) dt \tag{2.16}$$

In other way, one may define as below:
If the kernel in Eq. (2.1) is $K(t, s) = e^{ist}$, then the integral transform is called as FT.

If $f(t)$ is an even function, then the FT is called as cosine FT and the same is given as

$$F_c(f(t)) = \int_0^\infty f(s)\cos(st)dt$$

Similarly, if $f(t)$ is an odd function, then we have sine FT and it is written as

$$F_s(f(t)) = \int_0^\infty f(s)\sin(st)dt$$

In this regard, FTs of some standard functions are presented in Table 2.2. Here, $F(s)$, $G(s)$, and $U(s, t)$ are the FTs of $f(x)$, $g(x)$, and $u(x, t)$ respectively.

2.3.1 Solution of Partial Differential Equations Using Fourier Transforms

Solution of partial differential equations using FTs [2] is briefly explained below.

Step (i): First, apply FT on both sides of the given partial differential equation.
Step (ii): Use the derivative property of FT to find the FT of partial derivative as below (see Table 2.2):

$$F(u_t(x, t)) = F(u_t) = U_t(s, t) = \frac{\partial}{\partial t}U(s, t)$$

and

$$F(u_{xx}(x, t)) = F(u_{xx}) = (is)^2 U(s, t) = -s^2 U(s, t),$$

Table 2.2 Fourier transforms.

S. no.	Function	Fourier transform
1	Delta function $\delta(x)$	1
2	Exponential function $e^{-a|x|}$	$\dfrac{2a}{a^2 + s^2}$
3	Gaussian $e^{-x^2/2\sigma^2}$	$\sigma\sqrt{2\pi}\,e^{-\sigma^2 s^2/2}$
4	Derivative $f'(x)$	$isF(s)$
5	Convolution property: $f(x) * g(x)^{a)}$	$F(s)G(s)$
6	nth derivative $\dfrac{d^n f}{dt^n}$	$(2\pi is)^n F(s)$
7	nth partial derivative $\dfrac{\partial^n}{\partial x^n}u(x, t)$	$(is)^n U(s, t)$
8	nth partial derivative $\dfrac{\partial^n}{\partial t^n}u(x, t)$	$\cdot\ \dfrac{\partial^n}{\partial t^n}U(s, t)$

a) Convolution of $f(x)$ and $g(x)$ is $f(x) * g(x) = \int_{-\infty}^{\infty} f(\tau)g(x - \tau)d\tau$.

then simplify algebraically to obtain ordinary differential equation with independent variable t.

Step (iii): Solve the above obtained ordinary differential equation to get $U(s, t)$.

Step (iv): Find the inverse FT of $U(s, t)$ to obtain $u(x, t)$ which is the solution of the given differential equation.

Example 2.4   Consider one-dimensional heat equation

$$u_t - c u_{xx} = 0 \tag{2.17}$$

subject to the initial condition:

$$u(x, 0) = f(x) \tag{2.18}$$

We know that the FT [1, 2] of the function $u_t(x, t)$ and $u_{xx}(x, t)$ are

$$\left. \begin{array}{l} F(u_t(x, t)) = F(u_t) = \hat{u}_t(s, t) = \dfrac{\partial}{\partial t}\hat{u}(s, t) \text{ and} \\[2mm] F(u_{xx}(x, t)) = F(u_{xx}) = (is)^2 \hat{u}(s, t) = -s^2 \hat{u}(s, t) \end{array} \right\} \tag{2.19}$$

respectively. By applying FT on Eq. (2.17) and using Eq. (2.19), we get

$$\frac{\partial}{\partial t}\hat{u}(s, t) + cs^2 \hat{u}(s, t) = 0 \tag{2.20}$$

$$\hat{u}(s, 0) = \hat{f}(s).$$

Equation (2.20) is an ordinary differential equation with independent variable t, whose solution is

$$\hat{u}(s, t) = A(s)e^{-cs^2 t} \tag{2.21}$$

where $A(s)$ is an arbitrary function.

Using initial condition $\hat{u}(s, 0) = \hat{f}(s)$ in Eq. (2.21), we get $A(s) = \hat{f}(s)$. Therefore, the solution (2.20) may be written as

$$\hat{u}(s, t) = \hat{f}(s)e^{-cs^2 t} \tag{2.22}$$

Next, we use inverse FT along with convolution property (see Table 2.2) to find $u(x, t)$. For that, let us assume $g(x, t)$ is the inverse FT of $e^{-cs^2 t}$ (see Gaussian function transform in Table 2.2) and it may be found as

$$\psi(x, t) = F^{-1}(e^{-cs^2 t}) = F^{-1}\left(e^{-(\sqrt{2ct})^2\, s^2/2}\right)$$

$$= F^{-1}\left(e^{-(\sqrt{2ct})^2\, s^2/2}\right)$$

$$= \frac{1}{\sqrt{2ct}\sqrt{2\pi}}e^{-x^2/2(\sqrt{2ct})^2}$$

$$= \frac{1}{\sqrt{4\pi ct}} e^{-\frac{1}{4ct}x^2}$$

Thus, the final solution of the given PDE (2.17) for the function $f(x)$ may be written as

$$u(x, t) = (g(., t) * f(.))(x)$$

$$= \int_{-\infty}^{\infty} g(x - \tau, t) f(\tau) dy$$

$$= \frac{1}{\sqrt{4\pi ct}} \int_{-\infty}^{\infty} e^{-\frac{(x-\tau)^2}{4ct}} f(\tau) d\tau$$

For more details and applications of Laplace and Fourier transforms, one may go through Refs. [1–8] and therein.

Exercise

1   Use LT to solve the below initial value problems
    (a) $y'' + y = 2\cos(t)$, $y(0) = 3$, $y'(0) = 4$
    (b) $y'' + 9y = g(t)$, $y(0) = 0$, $y'(0) = 0$, where $g(t) = \begin{cases} 2, & 0 \leq t < 6 \\ 0, & t > 6 \end{cases}$
    (c) $y'' - y' - 2y = 12u_\pi(t)\sin(t)$, $y(0) = 1$, $y'(0) = -1$

2   Solve the given partial differential equation using LT $u_t + u_x = 0, x > 0, t > 0$ subject to the conditions $u(x, 0) = \sin(x)$, $u(0, t) = 0$.

References

1 Widder, D.V. (2015). *Laplace Transforms (PMS-6)*. Princeton, NJ: Princeton University Press.
2 Beerends, R.J., ter Morsche, H.G., van den Berg, J.C., and van de Vrie, E.M. (2003). *Fourier and Laplace Transforms*. Cambridge: Cambridge University Press Translated from the 1992 Dutch edition by Beerends.
3 Duffy, D.G. (2004). *Transform Methods for Solving Partial Differential Equations*. Boca Raton, FL: Chapman and Hall/CRC.
4 Bracewell, R.N. and Bracewell, R.N. (1986). *The Fourier Transform and Its Applications*, vol. 31999. New York: McGraw-Hill.
5 Haberman, R. (1983). *Elementary Applied Partial Differential Equations: With Fourier Series and Boundary Value Problems*. Upper Saddle River, NJ: Prentice Hall.
6 Spiegel, M.R. (1965). *Laplace Transforms*. New York: McGraw-Hill.

7 Marks, R.J. (2009). *Handbook of Fourier Analysis and Its Applications.* Oxford: Oxford University Press.

8 Poularikas, A.D. (2010). *Transforms and Applications Handbook.* Boca Raton, FL: CRC Press.

9 Kreyszig, E. (2010). *Advanced Engineering Mathematics.* New Delhi: Wiley.

3

Weighted Residual Methods

3.1 Introduction

Weighted residual is treated as another powerful method for computation of solution to differential equations subject to boundary conditions referred to as boundary value problems (BVPs). Weighted residual method (WRM) is an approximation technique in which solution of differential equation is approximated by linear combination of trial or shape functions having unknown coefficients. The approximate solution is then substituted in the governing differential equation resulting in error or residual. Finally, in the WRM the residual is forced to vanish at average points or made as small as possible depending on the weight function in order to find the unknown coefficients. WRMs, viz. collocation and Galerkin methods, have been discussed by Gerald and Wheatley [1]. Further discussion of various WRMs may be found in standard books viz. [2–4]. As regards, least-square method for solving BVPs has been given by Locker [5]. Weighted residual-based finite-element methods are discussed in Refs. [1–3, 6, 7]. In Chapter 6, finite-element discretization approach using Galerkin WRM has been introduced. Moreover, sometimes trial or shape functions taken as boundary characteristic orthogonal polynomials are advantageous. So, Chapter 4 is dedicated in solving BVP using boundary characteristic orthogonal polynomials incorporated into Galerkin and Rayleigh–Ritz methods.

In this regard, this chapter is organized such that various WRMs, viz. collocation, subdomain, least-square, and Galerkin methods applied for solving ordinary differential equations subject to boundary conditions have been illustrated in Sections 3.2, 3.3, 3.4, and 3.5 respectively. For better understanding of the methods, comparative results for specific differential equations with respect to WRMs are also included. Lastly, few exercise problems are also given at the end for self-validation.

Let us consider an ordinary differential equation,

$$L(u) = f(x) \tag{3.1}$$

Advanced Numerical and Semi-Analytical Methods for Differential Equations, First Edition.
Snehashish Chakraverty, Nisha Rani Mahato, Perumandla Karunakar, and Tharasi Dilleswar Rao.
© 2019 John Wiley & Sons, Inc. Published 2019 by John Wiley & Sons, Inc.

subject to boundary conditions (*BCs*) in the domain Ω. Here, L is the linear differential operator acting on u and f is the applied force. In WRM, the solution $u(x)$ of Eq. (3.1) is approximately considered as $\hat{u}(x) = \sum_{i=0}^{n} c_i \phi_i(x)$ satisfying the *BCs*. Here, c_i are the unknown coefficients yet to be determined for the trial functions $\phi_i(x)$ for $i = 0, 1, \ldots, n$, where $\phi_i(x)$ are linearly independent to each other. The assumed solution is substituted in the governing differential equation (3.1) resulting in error or residual. This residue is then minimized or forced to vanish in the domain Ω, resulting in system of algebraic equations in terms of unknown coefficients c_i. As such, WRMs mainly consist of the following steps:

Step (i): Assume an approximate solution

$$u(x) \approx \hat{u}(x) = \sum_{i=0}^{n} c_i \phi_i(x) \tag{3.2}$$

involving linearly independent trial functions $\phi_i(x)$ such that $\hat{u}(x)$ satisfies the boundary conditions. Alternatively,

$$\hat{u}(x) = u_0(x) + \sum_{i=0}^{n} c_i \phi_i(x) \tag{3.3}$$

may also be considered where $u_0(x)$ is the function satisfying the *BCs*. Generally, the choice of trial functions are considered such that the shape functions interpolate the desired solution subject to boundary conditions as mentioned in Eq. (3.2) or Eq. (3.3). Various cases studied for assumption of trial functions may be found in Ref. [2]. One possible assumption for the shape functions may be $\phi_0 = 1$, $\phi_1 = x$, and $\phi_i = x^{i-1}(x - X)$ for $i = 2, 3, \ldots, n - 1$ over the domain $\Omega = [0, X]$.

Step (ii): Substitute $\hat{u}(x)$ in the given differential equation $L(\hat{u}) - f(x) \neq 0$ that results in error.

Step (iii): The measure of error is considered as residual,

$$R(x) = L(\hat{u}) - f(x). \tag{3.4}$$

Step (iv): An arbitrary weight function $w_i(x)$ is then multiplied in Eq. (3.3) and integrated over Ω resulting in

$$\int_{\Omega} w_i(x)[L(\hat{u}) - f(x)]dx = \int_{\Omega} w_i(x)R(x)dx \neq 0$$

for $i = 0, 1, \ldots, n$. It is cumbersome to identically make the residue zero in the entire domain. So, the integral is either set to vanish at finite points or made as small as possible depending on the weight function.

Step (v): Forcing the integral to vanish over the entire domain Ω using

$$\int_\Omega w_i(x)R(x)dx = 0 \tag{3.5}$$

$n + 1$ independent algebraic equations are obtained using Eq. (3.5) for computing unknowns c_i, where $i = 0, 1, \ldots, n$.

It is worth mentioning that in Eq. (3.2), $u(x) = \hat{u}(x)$ as $n \to \infty$. Moreover, depending on weight function, there exist different types of WRMs, viz. collocation, least-square, Galerkin, and Rayleigh–Ritz methods. In this regard, WRMs are discussed in Sections 3.2–3.5.

3.2 Collocation Method

In the collocation method [1–3, 7], the weight function is taken in terms of Dirac delta function δ as

$$w_k(x) = \delta(x - x_k) = \begin{cases} 1, & x = x_k \\ 0, & \text{otherwise} \end{cases} \tag{3.6}$$

where $k = 0, 1, \ldots, n$. Often in literature studies, this approach is also referred to as the point collocation method. Now, using Eq. (3.6), the integrand

$$\int_\Omega w_k(x)R(x)dx = R(x_k).$$

Generally, the residual is set to zero at $n + 1$ distinct points x_k within the domain Ω,

$$R(x_k) = L\left(\sum_{i=0}^n c_i\phi_i(x_k)\right) - f(x_k) = 0 \Rightarrow \sum_{i=0}^n c_iL(\phi_i(x_k)) = f(x_k) \tag{3.7}$$

for computing the unknown coefficients c_i for $n - 1$ collocating points. For clear understanding, a BVP is solved using the collocation method in Example 3.1.

Example 3.1 Solve the ordinary differential equation,

$$u'' + 4u = x \tag{3.8}$$

subject to boundary conditions $u(0) = 1$ and $u\left(\frac{\pi}{4}\right) = 2$ within the domain $x \in \left[0, \frac{\pi}{4}\right]$ using the collocation method.

Solution Here, $L(u) = u'' + 4u$ and $f(x) = x$. Let us consider the approximate solution as given in Eq. (3.2) for $n = 3$ be $\hat{u}(x) = \hat{u}_3(x) = c_0\phi_0(x) + c_1\phi_1(x) + c_2\phi_2(x) + c_3\phi_3(x)$ having trial functions $\phi_0(x) = 1$, $\phi_1(x) = x$, $\phi_2(x) = x\left(x - \frac{\pi}{4}\right)$, and $\phi_3(x) = x^2\left(x - \frac{\pi}{4}\right)$. The approximate solution satisfies the boundary conditions that helps in computation of unknown coefficients as

$$\hat{u}(0) = c_0\phi_0(0) + c_1\phi_1(0) + c_2\phi_2(0) + c_3\phi_3(0) \Rightarrow c_0 = 1,$$

$$\hat{u}\left(\frac{\pi}{4}\right) = c_0\phi_0\left(\frac{\pi}{4}\right) + c_1\phi_1\left(\frac{\pi}{4}\right) + c_2\phi_2\left(\frac{\pi}{4}\right) + c_3\phi_3\left(\frac{\pi}{4}\right)$$

$$\Rightarrow 2 = c_0 + \frac{\pi}{4}c_1 \Rightarrow c_1 = \frac{4}{\pi}.$$

Accordingly, the approximate solution reduces to $\hat{u}(x) = 1 + \frac{4}{\pi}x + c_2x\left(x - \frac{\pi}{4}\right) + c_3x^2\left(x - \frac{\pi}{4}\right)$. The derivatives may be obtained as $\hat{u}'(x) = \frac{4}{\pi} + c_2\left(2x - \frac{\pi}{4}\right) + c_3x\left(3x - \frac{\pi}{2}\right)$ and $\hat{u}''(x) = 2c_2 + \left(6x - \frac{\pi}{2}\right)c_3$. Further, the residual reduces to

$$R(x_k) = \hat{u}''(x_k) + 4\hat{u}(x_k) - f(x_k)$$

$$\Rightarrow R(x_k) = (2 + 4x^2 - \pi x)c_2 + \left(4x^3 - \pi x^2 + 6x - \frac{\pi}{2}\right)c_3 + \left(4 - x + \frac{16x}{\pi}\right).$$

Now, the remaining two unknown coefficients c_2 and c_3 are computed for two collocating points x_1 and x_2 as given in Figure 3.1.

Then, the residual $R(x_k)$ is set to zero at the collocating points $x_1 = \frac{\pi}{12}$ and $x_2 = \frac{\pi}{6}$ resulting in algebraic equations:

$$R(x_1) = \left(2 - \frac{\pi^2}{18}\right)c_2 - \frac{\pi^3}{216}c_3 - \frac{\pi}{12} + \frac{16}{3} = 0 \tag{3.9a}$$

and

$$R(x_2) = \left(2 - \frac{\pi^2}{18}\right)c_2 + \left(\frac{\pi}{2} - \frac{\pi^3}{108}\right)c_3 - \frac{\pi}{6} + \frac{20}{3} = 0 \tag{3.9b}$$

Solving Eqs. (3.9a) and (3.9b), we obtain the approximate solution as

$$\hat{u}(x) = 1 + 4.0754x - 2.9781x^2 - 0.7508x^3.$$

In order to verify the correctness of the method, the approximate solution is compared with the exact solution $u(x) = \cos(2x) + \left(2 - \frac{\pi}{16}\right)\sin(2x) + \frac{x}{4}$ and incorporated into Figure 3.2.

Also, higher-order approximations, that is $\hat{u}_4(x) = \sum_{i=0}^{4} c_i\phi_i(x)$ and $\hat{u}_7(x) = \sum_{i=0}^{7} c_i\phi_i(x)$, are incorporated into Figure 3.2 that yield better accuracy.

Figure 3.1 Collocation points for the domain $\Omega = \left[0, \frac{\pi}{4}\right]$.

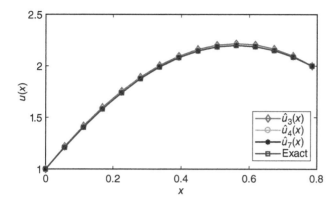

Figure 3.2 Comparison of the collocation method solution of $u'' + 4u = x$, $u(0) = 1$, $u\left(\frac{\pi}{4}\right) = 2$ with the exact solution.

3.3 Subdomain Method

In the subdomain method [2], the domain Ω is finitely divided into nonoverlapping n subdomains or subintervals $\Omega_{k+1} = [x_k, x_{k+1}]$, where $k = 0, 1, \ldots, n-1$. Here, the weight function is selected as unity in the subdomains, otherwise zero,

$$w_k(x) = \begin{cases} 1, & x \in \Omega_k \\ 0, & x \notin \Omega_k \end{cases} \tag{3.10}$$

Then, Eq. (3.10) reduces to

$$\int_{x_k}^{x_{k+1}} w_k(x)R(x)dx = \int_{x_k}^{x_{k+1}} R(x)dx = 0 \quad \text{for} \quad k = 0, 1, \ldots, n. \tag{3.11}$$

We rewrite Eq. (3.11) as $\int_{x_k}^{x_{k+1}} L\left(\sum_{i=0}^{n} c_i \phi_i(x)\right) - f(x)dx \Rightarrow \sum_{i=0}^{n} c_i \int_{x_k}^{x_{k+1}} L(\phi_i(x))$
$dx = \int_{x_k}^{x_{k+1}} f(x)dx$, which in turn is expressed in terms of system of equations:

$$\begin{bmatrix} \int_{x_0}^{x_1} L(\phi_0(x))dx & \int_{x_0}^{x_1} L(\phi_1(x))dx & \cdots & \int_{x_0}^{x_1} L(\phi_n(x))dx \\ \int_{x_1}^{x_2} L(\phi_0(x))dx & \int_{x_1}^{x_2} L(\phi_1(x))dx & \cdots & \int_{x_1}^{x_2} L(\phi_n(x))dx \\ \vdots & \vdots & \ddots & \vdots \\ \int_{x_{n-1}}^{x_n} L(\phi_0(x))dx & \int_{x_{n-1}}^{x_n} L(\phi_1(x))dx & \cdots & \int_{x_{n-1}}^{x_n} L(\phi_n(x))dx \end{bmatrix} \begin{Bmatrix} c_0 \\ c_1 \\ \vdots \\ c_n \end{Bmatrix} = \begin{Bmatrix} \int_{x_0}^{x_1} f(x)dx \\ \int_{x_1}^{x_2} f(x)dx \\ \vdots \\ \int_{x_{n-1}}^{x_n} f(x)dx \end{Bmatrix}$$

$$\tag{3.12}$$

The unknown coefficients in the approximate solution $\hat{u}(x) = \sum_{i=0}^{n} c_i \phi_i(x)$ are then obtained by solving Eq. (3.12). An example illustrating the usage of the subdomain method for solving BVPs is considered in Example 3.2.

Example 3.2 Solve the BVP,

$$y'' + x^2 = 0$$

subject to $y(0) = 2$ and $y(1) = 0$ within the domain $x \in [0, 1]$ using the subdomain method.

Solution Here, $L(y) = y''$ and $f(x) = -x^2$. Let us assume the approximate solution be $\hat{y}(x) = \hat{y}_n(x) = \sum_{i=0}^{n} c_i \phi_i(x)$ in terms of trial functions $\phi_0(x) = 1$, $\phi_1(x) = x$, and $\phi_i(x) = x^{i-1}(1 - x)$ for $i = 2, 3, \ldots, n$. For $n = 3$, the approximate solution satisfying BCs reduces to

$$\hat{y}(x) = 2 - 2x + c_2 \phi_2(x) + c_3 \phi_3(x).$$

Then, we obtain the residue as $R(x) = \hat{y}''(x) + x^2 = -2c_2 + 2c_3(1 - 3x) + x^2$. In order to compute the remaining two unknowns, the domain is divided into two subdomains $\Omega_1 = \left[0, \frac{1}{2}\right]$, $\Omega_2 = \left[\frac{1}{2}, 1\right]$ and the residue is forced to zero in the subdomains as

$$\int_{x_k}^{x_{k+1}} R(x)dx = 0 \Rightarrow \int_{x_k}^{x_{k+1}} (-2c_2 + 2c_3(1 - 3x) + x^2)dx = 0 \quad \text{for} \quad k = 0, 1.$$

We further obtain the equations for two subdomains,

$$\int_0^{\frac{1}{2}} (-2c_2 + 2c_3(1 - 3x) + x^2)dx = 0 \tag{3.13a}$$

$$\int_{\frac{1}{2}}^1 (-2c_2 + 2c_3(1 - 3x) + x^2)dx = 0 \tag{3.13b}$$

resulting in system of equations:

$$\begin{bmatrix} 1 & -\dfrac{1}{4} \\ 1 & \dfrac{5}{4} \end{bmatrix} \begin{Bmatrix} c_2 \\ c_3 \end{Bmatrix} = \begin{Bmatrix} \dfrac{1}{24} \\ \dfrac{7}{24} \end{Bmatrix}.$$

The approximate solution $\hat{y}(x) = 2 - 2x + \frac{x}{12}(1 - x) + \frac{x^2}{6}(1 - x) = 2 - \frac{23}{12}x - \frac{1}{12}x^2 - \frac{1}{6}x^3$ is now verified by comparing with the exact solution $y(x) = 2 - \frac{23}{12}x - \frac{1}{12}x^4$ in Figure 3.3.

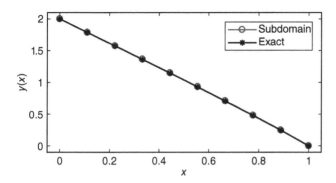

Figure 3.3 Comparison of solution of Example 3.2 using the subdomain method with the exact solution.

3.4 Least-square Method

In the least-square method [2, 3, 5], the residue given in Eq. (3.4) is squared and integrated over the entire domain Ω,

$$I = \int_{\Omega} R^2(x; c_j)dx. \tag{3.14}$$

The integrand I is then minimized using $\frac{\partial}{\partial c_k} \int_{\Omega} R^2(x; c_j)dx = 0$, where c_j are unknown coefficients of approximate solution $\hat{u}(x) = \sum_{i=0}^{n} c_j \phi_j(x)$, which further reduces to

$$2 \int_{\Omega} R(x; c_k)\frac{\partial R}{\partial c_k}dx = 0 \quad \text{for} \quad k = 0, 1, \dots, n$$

$$\Rightarrow \int_{\Omega} R(x; c_k)\frac{\partial R}{\partial c_k}dx = 0. \tag{3.15}$$

From Eq. (3.15), the weight function for the least-square method is considered as

$$w_k = \frac{\partial R}{\partial c_k} \quad \text{for} \quad k = 0, 1, \dots, n. \tag{3.16}$$

Example 3.3 Let us consider a BVP,

$$y'' - \cos(x) = 0$$

subject to $y(0) = 0$ and $y(1) = 0$ within the domain $x \in [0, 1]$. Solve the BVP using the least-square method.

Solution Assume the approximate solution be $\hat{y}(x) = \sum_{i=0}^{5} c_i \phi_i(x)$ in terms of trial functions $\phi_0(x) = 1$, $\phi_1(x) = x$, and $\phi_i(x) = x^{i-1}(1-x)$ for $i = 2, 3, ..., 5$. Using BCs, the values of c_0 and c_1 are obtained as $c_0 = c_1 = 0$. In order to compute the remaining unknown coefficients, the residue $R(x) = \hat{y}''(x) - \cos(x) = -2c_2 + 2c_3(1 - 3x) + 6c_4 x(1 - 2x) + 4c_5 x^2(3 - 5x)$ is minimized in the domain $[0, 1]$ resulting in equations:

$$\int_0^1 \left(R\frac{\partial R}{\partial c_2} \right) dx = 4c_2 + 2c_3 + 2c_4 + 2c_5 + 2\sin(1) = 0, \tag{3.17a}$$

$$\int_0^1 \left(R\frac{\partial R}{\partial c_3} \right) dx = 2c_2 + 4c_3 + 4c_4 + 4c_5 + 4\sin(1) + 6\cos(1) - 6 = 0, \tag{3.17b}$$

$$\int_0^1 \left(R\frac{\partial R}{\partial c_4} \right) dx = 2c_2 + 4c_3 + \frac{24}{5}c_4 + \frac{26}{5}c_5 - 18\sin(1) + 18\cos(1) + 6 = 0 \tag{3.17c}$$

and

$$\int_0^1 \left(R\frac{\partial R}{\partial c_5} \right) dx = 2c_2 + 4c_3 + \frac{26}{5}c_4 + \frac{208}{35}c_5 - 88\sin(1)$$
$$- 84\cos(1) + 120 = 0. \tag{3.17d}$$

On solving Eqs. (3.17a)–(3.17d), we get $c_2 = \cos(1) - 1$, $c_3 = 391\cos(1) + 178\sin(1) - 361$, $c_4 = 1119 - 592\sin(1) - 1149\cos(1)$, and $c_5 = 756\cos(1) + 413\sin(1) - 756$.

Accordingly, the approximate solution to four decimal places is obtained as $\hat{y}(x) = -0.4597x + 0.49997x^2 + 0.0018x^3 - 0.0458x^4 + 0.0039x^5$ and the comparative plot with the exact solution $y(x) = 1 - \cos(x) + (\cos(1) - 1)x$ is incorporated into Figure 3.4.

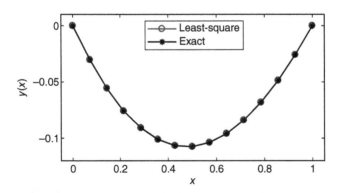

Figure 3.4 Comparison of solution of Example 3.3 using the least-square method with the exact solution.

3.5 Galerkin Method

In the Galerkin method [1–3, 7], the weight function is considered in terms of trial functions,

$$w_k = \phi_k(x) \quad \text{for} \quad k = 0, 1, \ldots, n \tag{3.18}$$

such that ϕ_k are $(n+1)$ basis functions of x satisfying the boundary conditions. As such, Eq. (3.5) reduces to

$$\int_\Omega w_k(x)R(x)dx = \int_\Omega \phi_k(x)R(x)dx = 0. \tag{3.19}$$

Then, the system of equations in terms of unknown coefficients are obtained using Eq. (3.19). To illustrate Galerkin's method, we consider an example problem given below.

Example 3.4   Solve the following differential equation

$$u'' = x$$

subject to boundary conditions $u(0) = 1$ and $u(2) = 2$ using the Galerkin method.

Solution   Let us assume the approximate solution be $\hat{u}(x) = \sum_{i=0}^{3} c_i\phi_i(x)$ having trial functions $\phi_0(x) = 1$, $\phi_1(x) = x$, $\phi_2(x) = x(2-x)$, and $\phi_3(x) = x^2(2-x)$. It is to be noted that the approximate solution satisfies the given boundary conditions. As such,

$$\hat{u}(0) = c_0 \Rightarrow c_0 = 1 \tag{3.20a}$$

and

$$\hat{u}(2) = c_0 + 2c_1 = 2 \Rightarrow c_1 = \frac{1}{2}. \tag{3.20b}$$

Next, the residue for the Galerkin method is obtained as

$$R(x) = \sum_{i=0}^{3} c_i\phi_i''(x) - x = -2c_2 + 2c_3(2 - 3x) - x$$

and accordingly, the remaining coefficients c_2 and c_3 are computed using

$$\int_0^2 \phi_2 R(x)dx = 0 \Rightarrow 2c_2 + 2c_3 + 1 = 0, \tag{3.21a}$$

$$\int_0^2 \phi_3 R(x)dx = 0 \Rightarrow 5c_2 + 8c_3 + 3 = 0. \tag{3.21b}$$

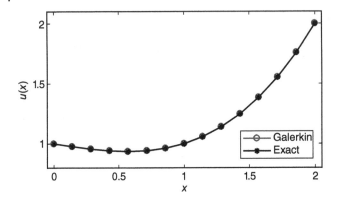

Figure 3.5 Comparison of solution of Example 3.4 using the Galerkin method with the exact solution.

Solving Eqs. (3.21a) and (3.21b), we get $c_2 = -\frac{1}{3}, c_3 = -\frac{1}{6}$ and the approximate solution reduces to

$$\hat{u}(x) = 1 + \frac{x}{2} - \frac{x(2-x)}{3} - \frac{x^2(2-x)}{6}. \tag{3.22}$$

Then, the comparison of solution obtained in Eq. (3.22) is done with the exact solution $u(x) = 1 - \frac{x}{6} + \frac{x^3}{6}$ in Figure 3.5.

It is to be noted that if the choice of trial functions is considered such that they form orthogonal basis, then the Galerkin method is found to be more efficient as from Eq. (3.19) we have then $\phi_i(x) \cdot \phi_j(x) = \begin{cases} \phi_i^2(x), & i = j \\ 0, & i \neq j \end{cases}$. Such approach is considered in Chapter 4 in terms of boundary characteristic orthogonal polynomials.

3.6 Comparison of WRMs

In this section, we check the efficiency of various WRMs discussed in Sections 3.2–3.5 by comparing the solution obtained using collocation, subdomain, least-square, and Galerkin methods.

Example 3.5 Compare approximate solutions obtained using collocation, subdomain, least-square, and Galerkin methods for the BVP,

$$u'' + u' + u = x \tag{3.23}$$

subject to $u(0) = 0$ and $u(1) = 0$.

Solution The exact solution of Eq. (3.23) is

$$u(x) = e^{-\frac{x}{2}} \left(\cos\left(\frac{\sqrt{3}}{2}x \right) + \sin\left(\frac{\sqrt{3}}{2}x \right) \cot\left(\frac{\sqrt{3}}{2} \right) + x - 1 \right).$$

Let us consider a cubic approximate solution $\hat{u}(x) = c_0\phi_0(x) + c_1\phi_1(x) + c_2\phi_2(x) + c_3\phi_3(x)$ having trial functions $\phi_0(x) = 1$, $\phi_1(x) = x$, $\phi_2(x) = x(1-x)$, and $\phi_3(x) = x^2(1-x)$. Using the given boundary conditions, we get

$$\hat{u}(0) = c_0\phi_0(0) + c_1\phi_1(0) + c_2\phi_2(0) + c_3\phi_3(0) = 0 \Rightarrow c_0 = 0, \quad (3.24a)$$

$$\hat{u}(1) = c_0\phi_0(1) + c_1\phi_1(1) + c_2\phi_2(1) + c_3\phi_3(1) = 0 \Rightarrow c_1 = 0. \quad (3.24b)$$

Then, the approximate result reduces to $\hat{u}(x) = c_2x(1-x) + c_3x^2(1-x)$. As such, the residual is obtained as

$$R(x) = (2 + 4x^2 - \pi x)c_2 + \left(4x^3 - \pi x^2 + 6x - \frac{\pi}{2} \right)c_3 + \left(4 - x + \frac{16x}{\pi} \right).$$

$$(3.25)$$

Now, the remaining two unknown coefficients c_2 and c_3 are computed using WRMs resulting in solutions,

1. Collocation method: $\hat{u}(x) = -0.2515x + 0.1781x^2 + 0.0733x^3$
2. Subdomain method: $\hat{u}(x) = -0.2340x + 0.1567x^2 + 0.0774x^3$
3. Least-square method: $\hat{u}(x) = -0.2357x + 0.1615x^2 + 0.0742x^3$
4. Galerkin method: $\hat{u}(x) = -0.2447x + 0.1702x^2 + 0.0745x^3$

Then, the approximate solutions are compared with the exact solution in Figure 3.6.

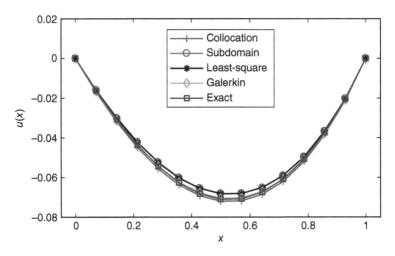

Figure 3.6 Comparison of the approximate solution of Example 3.5 with the exact solution.

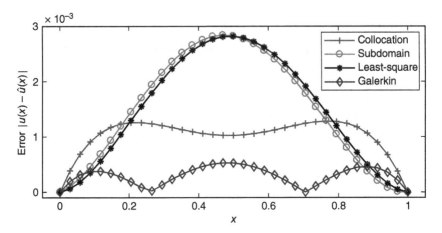

Figure 3.7 Error plot of approximate solutions for Example 3.5.

Further, error $= |u(x) - \hat{u}(x)|$ is computed and the respective errors for WRMs are depicted in Figure 3.7.

It is clearly seen from Figures 3.6 and 3.7 that the Galerkin method provides the best accuracy for Example 3.5 with the exact solution as the error is minimum compared to collocation, subdomain, and least-square methods.

Exercise

1   Apply the subdomain method to solve differential equation $y'' + y' + y = t^3$ subject to boundary conditions $y(0) = 1$ and $y(1) = 0$.

2   Use collocation and Galerkin methods to compute the approximate solution of following BVPs subject to boundary conditions $y(0) = 0$ and $y(1) = 0$ in the domain $\Omega = [0, 1]$.
    (a) $y'' + 2y = x^2$
    (b) $y'' + y = \sin(x)$.

3   Find the approximate solution of BVP $y'' - 2y' + x = 0$ subject to $y(0) = 1$ and $y(1) = 0$ using the least-square method.

References

1 Gerald, C.F. and Wheatley, P.O. (2004). *Applied Numerical Analysis*, 7e. New Delhi: Pearson Education Inc.

2 Finlayson, B.A. (2013). *The Method of Weighted Residuals and Variational Principles*, vol. 73. Philadelphia: SIAM.

3 Hatami, M. (2017). *Weighted Residual Methods: Principles, Modifications and Applications*. London: Academic Press.

4 Baluch, M.H., Mohsen, M.F.N., and Ali, A.I. (1983). Method of weighted residuals as applied to nonlinear differential equations. *Applied Mathematical Modelling* 7 (5): 362–365.

5 Locker, J. (1971). The method of least squares for boundary value problems. *Transactions of the American Mathematical Society* 154: 57–68.

6 Lindgren, L.E., 2009. From Weighted Residual Methods to Finite Element Methods. *Technical report*.

7 Logan, D.L. (2011). *A First Course in the Finite Element Method*, 5e. Stamford: Cengage Learning.

4

Boundary Characteristics Orthogonal Polynomials

4.1 Introduction

Boundary characteristic orthogonal polynomials (BCOPs) proposed by Bhat [1, 2] in 1985 have been used in various science and engineering problems. Further, many authors like Bhat and Chakraverty [3], Singh and Chakraverty [4] have also used BCOPs in different problems. BCOPs are found to be advantageous in well-known methods like Rayleigh–Ritz, Galerkin, collocation, etc.

BCOPs may be generated by using the Gram–Schmidt orthogonalization procedure [5]. The generated BCOPs have to satisfy the boundary conditions of the considered problem [3, 6]. Initially, the general approximated solution of the considered problem is assumed as linear combination of BCOPs. By substituting the approximated solution in the boundary value problem, one may get the residual [4, 7]. Further, by using the residual and following the algorithm of a particular method, one may develop a linear system of equations. In the final step, one can handle the obtained linear system by using any known analytical/numerical procedure. The orthogonal nature of the BCOPs makes the analysis simple and straightforward.

The following section presents the Gram–Schmidt orthogonalization process for generating orthogonal polynomials.

4.2 Gram–Schmidt Orthogonalization Process

Let us suppose a set of functions $(g_i(x), i = 1, 2, \ldots)$ in $[a, b]$. From these set of functions, one can construct appropriate orthogonal functions by using the well-known procedure known as the Gram–Schmidt orthogonalization [3, 5] process as follows:

$$\phi_1 = g_1,$$
$$\phi_2 = g_2 - \alpha_{21}\phi_1,$$

Advanced Numerical and Semi-Analytical Methods for Differential Equations, First Edition.
Snehashish Chakraverty, Nisha Rani Mahato, Perumandla Karunakar, and Tharasi Dilleswar Rao.
© 2019 John Wiley & Sons, Inc. Published 2019 by John Wiley & Sons, Inc.

$$\phi_3 = g_3 - \alpha_{31}\phi_1 - \alpha_{32}\phi_2,$$

$$\vdots$$

where $\alpha_{21} = \frac{\langle g_2, \phi_1 \rangle}{\langle \phi_1, \phi_1 \rangle}$, $\alpha_{31} = \frac{\langle g_3, \phi_1 \rangle}{\langle \phi_1, \phi_1 \rangle}$, $\alpha_{32} = \frac{\langle g_3, \phi_2 \rangle}{\langle \phi_2, \phi_2 \rangle}$, etc. and $\langle \ \rangle$ defines the inner product of the respective polynomials.

In general, one can write the above procedure as

$$\phi_1 = g_1,$$

$$\phi_i = g_i - \sum_{j=1}^{i-1} \alpha_{ij}\phi_j.$$

where $\alpha_{ij} = \frac{\langle g_i, \phi_j \rangle}{\langle \phi_j, \phi_j \rangle} = \frac{\int_a^b w(x)g_i(x)\phi_j(x)dx}{\int_a^b w(x)\phi_j(x)\phi_j(x)dx}$ and $w(x)$ is a weight function.

Next, we address the generation procedure for BCOPs.

4.3 Generation of BCOPs

The first member of BCOPs set viz. $\phi_0(x)$ is chosen as the simplest polynomial of the least order which satisfies the boundary conditions of the considered problem. The other members of the orthogonal set in the interval $a \leq x \leq b$ are generated by using the Gram–Schmidt process [3, 4] as follows:

$$\phi_1(x) = (x - l_1)\phi_0(x),$$

$$\vdots$$

$$\phi_k(x) = (x - l_k)\phi_{k-1}(x) - m_k\phi_{k-2}(x),$$

where $l_k = \frac{\int_a^b x\phi_{k-1}^2(x)dx}{\int_a^b \phi_{k-1}^2(x)dx}$ and $m_k = \frac{\int_a^b x\phi_{k-1}(x)\phi_{k-2}(x)dx}{\int_a^b \phi_{k-2}^2(x)dx}$.

Here, we consider $w(x) = 1$. The polynomials $\phi_k(x)$ satisfy the orthogonality condition:

$$\int_a^b \phi_i(x)\phi_j(x)dx = 0 \quad \text{if} \quad i \neq j$$

Next, we present Galerkin's method with BCOPs to handle various problems.

4.4 Galerkin's Method with BCOPs

Let us consider a second-order boundary value problem [4] on $[a, b]$ as

$$y'' + p(x)y = q(x), \quad y(a) = \alpha_1, \quad y(b) = \alpha_2 \tag{4.1}$$

An approximate solution of Eq. (4.1) is considered as

$$y(x) = \sum_{i=0}^{n} c_i \phi_i(x) \tag{4.2}$$

where $\phi_i(x)$ are BCOPs which satisfy the given boundary conditions and c_0, $c_1, \ldots, c_n$ are real constants.

By substituting Eq. (4.2) in Eq. (4.1), one may obtain the residual R as

$$R(x; c_0, c_1, \ldots, c_n) = \sum_{i=0}^{n} c_i \phi_i''(x) + p(x) \sum_{i=0}^{n} c_i \phi_i(x) - q(x) \tag{4.3}$$

The residual R is orthogonalized to the $(n+1)$ BCOPs $\phi_0, \phi_1, \ldots, \phi_n$ as

$$\int_a^b R(x; c_0, c_1, \ldots, c_n) \phi_j(x) dx = 0, \quad j = 0, \ldots, n \tag{4.4}$$

where $\int_a^b \phi_i(x) \phi_j(x) dx = 0, \quad i \neq j$.

From Eq. (4.4), one may obtain $(n+1)$ system of linear equations with $(n+1)$ unknowns which can be solved by any standard analytical method. Further, by substituting the evaluated constants $c_0, c_1, \ldots, c_n$ in Eq. (4.2) we get the approximate solution of the boundary value problem (4.1). One may note that the terms containing $\phi_0, \phi_1, \ldots, \phi_n$ will vanish due to the orthogonal property. This makes the method efficient.

Let us now solve a boundary value problem by using BCOPs with Galerkin's method.

Example 4.1 Find the approximate solution of $y'' + 2y + 4x = 0$ subject to the boundary conditions $y(0) = y(1) = 0$ by using Galerkin's method [3, 4] with BCOPs.

Solution The first two BCOPs generated (Section 4.3) by using the Gram–Schmidt orthogonalization are

$$\phi_0(x) = x(1-x) \quad \text{and} \quad \phi_1(x) = x^2(1-x) - \frac{1}{2}x(1-x).$$

Accordingly, a two-term approximation of the given equation is assumed as

$$y(x) = c_0 \phi_0(x) + c_1 \phi_1(x). \tag{4.5}$$

By substituting Eq. (4.5) in the given equation, one may obtain the residual R as

$$R(x; c_0, c_1) = c_0(\phi_0''(x) + 2\phi_0(x)) + c_1(\phi_1''(x) + 2\phi_1(x)) + 4x. \tag{4.6}$$

The residual R is then orthogonalized in terms of functions ϕ_0 and ϕ_1 as

$$\int_0^1 R(x; c_0, c_1)\phi_0(x)dx = 0, \tag{4.7}$$

$$\int_0^1 R(x; c_0, c_1)\phi_1(x)dx = 0, \tag{4.8}$$

where $\int_0^1 \phi_0(x)\phi_1(x)dx = 0$.

By solving the linear equations (4.7) and (4.8), we can evaluate the involved constants as

$$c_0 = \frac{5}{4} \quad \text{and} \quad c_1 = \frac{7}{10}$$

From Eq. (4.5) we may represent the two-term approximate solution of the given equation as

$$y(x) = \frac{5}{4}(x - x^2) + \frac{7}{10}\left(x^2(1-x) - \frac{1}{2}x(1-x)\right) \tag{4.9}$$

In Section 4.5, another approach viz. Rayleigh–Ritz method has been presented.

4.5 Rayleigh–Ritz Method with BCOPs

Let us consider a second-order boundary value problem [2, 4] as

$$y'' + P(x)y = R(x), \quad y(a) = a_1, \quad y(b) = b_1 \tag{4.10}$$

Then, one can find a functional (function of functions) $S(x, y, y')$ as

$$I = \int_a^b S(x, y, y')dx \tag{4.11}$$

where

$$S(x, y, y') = \left(\frac{dy}{dx}\right)^2 - Py^2 + 2Ry. \tag{4.12}$$

Now, from Eq. (4.12), one can have the differential equation (4.10) by applying the well-known Euler–Lagrange equation [8] as

$$\frac{\partial S}{\partial y} - \frac{d}{dx}\left(\frac{\partial S}{\partial y'}\right) = 0 \tag{4.13}$$

$$\Rightarrow -2Py + 2R - \frac{d}{dx}\left(2\frac{dy}{dx}\right) = 0 \Rightarrow y'' + P(x)y = R(x) \tag{4.14}$$

Further, assume an approximate solution with BCOPs to Eq. (4.10) as

$$y(x) = \sum_{i=0}^{n} c_i \phi_i(x) \tag{4.15}$$

where ϕ_i's are linearly independent BCOPs.

Let us substitute Eq. (4.15) in Eq. (4.11) and thus the integral "I" is evaluated as a function of c_i. The necessary conditions for the extremum value of "I" from ordinary calculus gives

$$\frac{\partial I}{\partial c_i} = 0, \quad i = 0, 1, \ldots, n \tag{4.16}$$

From Eq. (4.16), one may obtain $(n + 1)$ system of linear equations with $(n + 1)$ unknowns. The obtained system of equations (4.16) may be solved to evaluate c_i by using the standard analytical/numerical method to get the approximate solution of Eq. (4.10).

Next, we solve a boundary value problem by using BCOPs with the Rayleigh–Ritz method.

Example 4.2 Approximate the solution of $y'' = x + y$ subject to the boundary conditions $y(0) = y(1) = 0$ by using BCOPs in the Rayleigh–Ritz method.

Solution Given that

$$y'' = x + y, \quad y(0) = y(1) = 0 \tag{4.17}$$

Then the functional $S(x, y, y')$ is found as

$$S(x, y, y') = \left(\frac{dy}{dx}\right)^2 + y^2 + 2xy \tag{4.18}$$

and

$$I = \int_0^1 \left(\left(\frac{dy}{dx}\right)^2 + y^2 + 2xy \right) dx \tag{4.19}$$

Assuming a two-term approximate solution with BCOPs for Eq. (4.17), we have

$$y(x) = c_0 \phi_0(x) + c_1 \phi_1(x) \tag{4.20}$$

where $\phi_0(x) = x(1 - x)$ and $\phi_1(x) = x^2(1 - x) - \frac{1}{2}x(1 - x)$.

By substituting Eq. (4.20) in Eq. (4.19), one may obtain

$$I = \int_0^1 ((c_0 \phi_0'(x) + c_1 \phi_1'(x))^2 + (c_0 \phi_0(x) + c_1 \phi_1(x))^2 + 2x(c_0 \phi_0(x) + c_1 \phi_1(x))) dx \tag{4.21}$$

By applying necessary conditions for the extreme value of "I," we obtain

$$\frac{\partial I}{\partial c_0} = c_0 \int_0^1 ((\phi_0')^2 + \phi_0^2) dx + c_1 \int_0^1 ((\phi_0'\phi_1') + \phi_0\phi_1) dx + \int_0^1 x\phi_0 dx = 0$$

(4.22)

$$\frac{\partial I}{\partial c_1} = c_0 \int_0^1 ((\phi_0'\phi_1') + \phi_0\phi_1) dx + c_1 \int_0^1 ((\phi_1')^2 + \phi_1^2) dx + \int_0^1 x\phi_1 dx = 0$$

(4.23)

where $\int_0^1 \phi_0(x)\phi_1(x) dx = 0$ and $\int_0^1 (\phi_0'\phi_1') dx = 0$.

Further, by solving the linear equations (4.22) and (4.23), we may get

$$c_0 = -\frac{1}{6} \quad \text{and} \quad c_1 = -\frac{7}{43}.$$

The approximate solution of the considered problem (4.17) may be written as

$$y(x) = -\frac{1}{6}(x(1-x)) - \frac{7}{43}\left(x^2(1-x) - \frac{1}{2}x(1-x)\right)$$

(4.24)

Below, we solve a practical problem related to transverse vibration of a string.

Example 4.3   Solve the problem of transverse vibration of a string $T\frac{d}{dx}\left(\frac{du(x,t)}{dx}\right) = m\frac{d^2 u(x,t)}{dt^2}$, $u(0, t) = 0$, and $u(1, t) = 0$ by using BCOPs in the Rayleigh–Ritz method [2, 6].

Solution The corresponding differential equation of transverse vibration of a string may be represented as

$$T\frac{d}{dx}\left(\frac{du(x,t)}{dx}\right) = m\frac{d^2 u(x,t)}{dt^2}, \quad u(0,t) = 0 \quad \text{and} \quad u(1,t) = 0 \quad (4.25)$$

where

T = String tension,
m = Mass per unit length of the string.

By considering $u(x, t) = U(x)\cos(\omega t)$, for simple harmonic motion, where ω = system frequency, Eq. (4.25) becomes

$$\frac{d^2 U}{dx^2} + K^2 U = 0, \quad U(0) = 0 = U(1)$$

(4.26)

where $K^2 = \frac{\omega^2 m}{T}$.

Now, in order to apply the Rayleigh–Ritz method to Eq. (4.26), let us choose the functional:

$$S(x, U, U') = (U')^2 - K^2 U^2$$

(4.27)

We assume an approximate solution of Eq. (4.26) as

$$U(x) = \sum_{i=0}^{n} c_i \phi_i(x), \quad i = 0, 1, \ldots, n \tag{4.28}$$

where ϕ_i represents BCOPs.

By applying integration to Eq. (4.26), one may find "I" as

$$I = \int_0^1 ((U')^2 - K^2 U^2) dx. \tag{4.29}$$

Substituting the approximate solution (4.28) in Eq. (4.29), one may obtain

$$I = \int_0^1 \left[\left(\sum_{i=0}^{n} c_i \phi_i'(x) \right)^2 - K^2 \left(\sum_{i=0}^{n} c_i \phi_i(x) \right)^2 \right] dx \tag{4.30}$$

By finding the extremum values of I, we have

$$\frac{\partial I}{\partial c_i} = 0 \Rightarrow \sum_{i=0}^{n} c_i \left[\int_0^1 \phi_i' \phi_j' dx - K^2 \int_0^1 \phi_i \phi_j dx \right] = 0, \quad j = 0, 1, \ldots, n \tag{4.31}$$

where $\int_0^1 \phi_i' \phi_j' dx = 0$ and $\int_0^1 \phi_i \phi_j dx$ if $i \neq j$.

From Eq. (4.31), one can obtain $(n+1)$ system of linear equations with $(n+1)$ unknowns. Then, by applying any standard analytical/numerical methods we may compute the values of c_i. By substituting the values of c_i in Eq. (4.28), we may finally get the approximated solution of Eq. (4.25).

Exercise

1 Use BCOPs in Galerkin and Rayleigh–Ritz methods to approximate the solutions of the following differential equations:
   (i) $2y'' = 5x + 3$ subject to $y(0) = 1$, $y(1) = 3$.
   (ii) $2y'' - xy' + 3 = 0$ subject to $y(0) = 0$, $y'(2) = 4$.

References

1 Bhat, R.B. (1985). Natural frequencies of rectangular plates using characteristic orthogonal polynomials in Rayleigh-Ritz method. *Journal of Sound and Vibration* 102 (4): 493–499.
2 Bhat, R.B. (1986). Transverse vibrations of a rotating uniform cantilever beam with tip mass as predicted by using beam characteristic orthogonal polynomials in the Rayleigh-Ritz method. *Journal of Sound and Vibration* 105 (2): 199–210.

3 Bhat, R.B. and Chakraverty, S. (2004). *Numerical Analysis in Engineering*. Oxford: Alpha Science Int'l Ltd.

4 Singh, B. and Chakraverty, S. (1994). Boundary characteristic orthogonal polynomials in numerical approximation. *Communications in Numerical Methods in Engineering* 10 (12): 1027–1043.

5 Johnson, P.S. (2014). Gram-Schmidt Orthogonalization Process.

6 Singh, B. and Chakraverty, S. (1994). Flexural vibration of skew plates using boundary characteristic orthogonal polynomials in two variables. *Journal of Sound and Vibration* 173 (2): 157–178.

7 Chakraverty, S., Saini, H., and Panigrahi, S.K. (2008). Prediction of product parameters of fly ash cement bricks using two dimensional orthogonal polynomials in the regression analysis. *Computers and Concrete* 5 (5): 449–459.

8 Agrawal, O.P. (2002). Formulation of Euler–Lagrange equations for fractional variational problems. *Journal of Mathematical Analysis and Applications* 272 (1): 368–379.

5

Finite Difference Method

5.1 Introduction

Finite difference methods (FDM) are well-known numerical methods to solve differential equations by approximating the derivatives using different difference schemes [1, 2]. The finite difference approximations for derivatives are one of the simplest and oldest methods to solve differential equations [3]. Finite difference techniques in numerical applications began in the early 1950s as represented in Refs. [4, 5], and their advancement was accelerated by the emergence of computers that offered a convenient framework for dealing with complex problems of science and technology. Theoretical results have been found during the last five decades related to accuracy, stability, and convergence of the finite difference schemes (FDS) for differential equations. Many science and engineering models involve nonlinear and nonhomogeneous differential equations, and solutions of these equations are sometimes beyond the reach by analytical methods. In such cases, FDM may be found to be practical, particularly for regular domains.

There are various types and ways of FDS [6, 7] depending on the type of differential equations, stability, and convergence [8]. However, here only the basic idea of the titled method is discussed.

5.2 Finite Difference Schemes

In this section, the fundamental concept and FDS are addressed. The numerical solutions of differential equations based on finite difference provide us with the values at discrete grid points. Let us consider a domain (Figure 5.1) in the xy-plane. The spacing of the grid points in the x- and y-directions are assumed to be uniform and given by Δx and Δy. It is not necessary that Δx and Δy be uniform or equal to each other [1, 5, 8] always.

The grid points in the x-direction are represented by index i and similarly j in the y-direction. The concept of FDM is to replace the derivatives involved in

Advanced Numerical and Semi-Analytical Methods for Differential Equations, First Edition.
Snehashish Chakraverty, Nisha Rani Mahato, Perumandla Karunakar, and Tharasi Dilleswar Rao.
© 2019 John Wiley & Sons, Inc. Published 2019 by John Wiley & Sons, Inc.

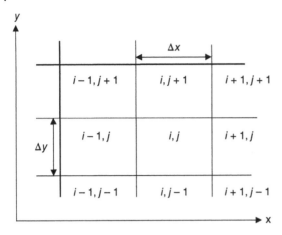

y

Figure 5.1 Discrete grid points.

the governing differential equations with algebraic difference techniques. This results in a system of algebraic equations which can be solved by any standard analytical/numerical methods for evaluating the dependent variables at the discrete grid points. There are three ways to express derivatives in FDS which are

1. Forward difference scheme
2. Backward difference scheme
3. Central difference scheme.

5.2.1 Finite Difference Schemes for Ordinary Differential Equations

5.2.1.1 Forward Difference Scheme
In this scheme, the rate of change of the function with respect to the variable x is represented between the current value at $x = x_i$ and the forward step at $x_{i+1} = x_i + \Delta x_i$. Mathematical representation of the derivative of the function $y(x)$ can be presented as

$$\Delta y_i = \frac{dy(x)}{dx}\bigg|_{x=x_i} \approx \frac{y_{i+1} - y_i}{x_{i+1} - x_i} = \frac{y_{i+1} - y_i}{\Delta x} = \frac{y_{i+1} - y_i}{h},$$

$$\Delta y_{i+1} = \frac{y_{i+2} - y_{i+1}}{h}, \quad \Delta y_{i+2} = \frac{y_{i+3} - y_{i+2}}{h}, \quad \text{etc.}$$

where h is the increment in step size.

The second-order derivative of the function at x can be derived as

$$\Delta^2 y_i = \frac{d}{dx}\left(\frac{dy(x)}{dx}\right)\bigg|_{x=x_i} = \frac{\Delta y_{i+1} - \Delta y_i}{h} = \frac{y_{i+2} - 2y_{i+1} + y_i}{h^2}$$

5.2.1.2 Backward Difference Scheme

In this case, we evaluate the rate of change of the function values between the current step at x_i and back step at $x_{i-1} = x_i - \Delta x$. Mathematically, we can represent it as

$$\nabla y_i = \frac{dy(x)}{dx}\bigg|_{x=x_i} \approx \frac{y_i - y_{i-1}}{h}$$

and the second-order derivative in the form

$$\nabla^2 y_i = \frac{d}{dx}\left(\frac{dy(x)}{dx}\right)\bigg|_{x=x_i} = \frac{\nabla y_i - \nabla y_{i-1}}{h} = \frac{y_i - 2y_{i-1} + y_{i-2}}{h^2}.$$

5.2.1.3 Central Difference Scheme

In the central difference scheme the rate of change of $y(x)$ is considered between the step at $x_i - \Delta x$ and the step ahead of x_i, i.e. $x_i + \Delta x$. Mathematically,

$$\Delta y_i \approx \frac{y_{i+1} - y_{i-1}}{x_{i+1} - x_{i-1}} = \frac{y_{i+1} - y_{i-1}}{2h}$$

and the second-order central difference form for the derivative is given by

$$\frac{d^2 y}{dx^2} \approx \frac{y_{i+1} - 2y_i + y_{i-1}}{h^2}.$$

5.2.2 Finite Difference Schemes for Partial Differential Equations

Similar to the FDS for ODE discussed in Section 5.2.1, FDS for Partial differential equations (PDEs) are given in Table 5.1.

To evaluate the derivative term $\frac{\partial^2 z}{\partial x \partial y}$, let us write the x-direction as a central difference of y-derivatives [9], and further we make use of central difference to find out the y-derivatives. Thus, we obtain

$$\frac{\partial^2 z}{\partial x \partial y} = \frac{\partial}{\partial x}\left(\frac{\partial z}{\partial y}\right) = \frac{\left(\frac{\partial z}{\partial y}\right)_{i+1,j} - \left(\frac{\partial z}{\partial y}\right)_{i-1,j}}{2(\Delta x)}$$

$$\cong \frac{1}{4\Delta x \Delta y}(z_{i+1,j+1} + z_{i-1,j-1} - z_{i+1,j-1} - z_{i-1,j+1}).$$

Similarly, other difference schemes may also be written.

5.3 Explicit and Implicit Finite Difference Schemes

Explicit and implicit approaches are used to obtain the numerical approximations to the time-dependent differential equations, as required for computer simulations [10, 11].

Table 5.1 Finite difference schemes for PDEs.

Scheme	Derivative approximation of $z(x, y)$	
	Derivative with respect to x	Derivative with respect to y
Forward	$\dfrac{\partial z}{\partial x} \approx \dfrac{z_{i+1,j} - z_{i,j}}{\Delta x}$	$\dfrac{\partial z}{\partial y} \approx \dfrac{z_{i,j+1} - z_{i,j}}{\Delta y}$
	$\dfrac{\partial^2 z}{\partial x^2} \approx \dfrac{z_{i+2,j} - 2z_{i+1,j} + z_{i,j}}{(\Delta x)^2}$	$\dfrac{\partial^2 z}{\partial y^2} \approx \dfrac{z_{i,j+2} - 2z_{i,j+1} + z_{i,j}}{(\Delta y)^2}$
Center	$\dfrac{\partial z}{\partial x} \approx \dfrac{z_{i+1,j} - z_{i-1,j}}{2\Delta x}$	$\dfrac{\partial z}{\partial y} \approx \dfrac{z_{i,j+1} - z_{i,j-1}}{2\Delta y}$
	$\dfrac{\partial^2 z}{\partial x^2} \approx \dfrac{z_{i+1,j} - 2z_{i,j} + z_{i-1,j}}{(\Delta x)^2}$	$\dfrac{\partial^2 z}{\partial y^2} \approx \dfrac{z_{i,j+1} - 2z_{i,j} + z_{i,j-1}}{(\Delta y)^2}$
Backward	$\dfrac{\partial z}{\partial x} \approx \dfrac{z_{i,j} - z_{i-1,j}}{\Delta x}$	$\dfrac{\partial z}{\partial y} \approx \dfrac{z_{i,j} - z_{i,j-1}}{\Delta y}$
	$\dfrac{\partial^2 z}{\partial x^2} \approx \dfrac{z_{i,j} - 2z_{i-1,j} + z_{i-2,j}}{(\Delta x)^2}$	$\dfrac{\partial^2 z}{\partial y^2} \approx \dfrac{z_{i,j} - 2z_{i,j-1} + z_{i,j-2}}{(\Delta y)^2}$

5.3.1 Explicit Finite Difference Method

Explicit finite difference scheme evaluates the state of a system at later time by using the known values at current time. Mathematically, the temperature at time $t + \Delta t$ depends explicitly on the temperature at time t and this scheme is relatively simple and computationally fast.

Let us introduce a parabolic PDE, namely heat equation, to understand the procedure for the explicit finite difference method.

Let us consider a time-dependent heat equation

$$\frac{\partial u(x, t)}{\partial t} = k \frac{\partial^2 u(x, t)}{\partial x^2}, \quad 0 \le x \le L, \quad t \ge 0 \tag{5.1}$$

with respect to initial condition

$$u(x, 0) = u_0(x),$$

and boundary conditions

$$u(0, t) = u_1(t), \quad u(L, t) = u_2(t).$$

Let the domain from $x = 0$ to $x = L$ be subdivided into N subparts so that $x = 0$ corresponds to $i = 0$ and $x = L$ corresponds to $i = N$ with $t = 0$ corresponding to $j = 0$. Then the initial conditions may be represented as

$$u(x, 0) = u_0(x) \Rightarrow u_{i,0} = u_0.$$

and the boundary conditions

$$u(0, t) = u_1(t) \Rightarrow u_{0,j} = u_1,$$
$$u(L, t) = u_2(t) \Rightarrow u_{N,j} = u_2.$$

The central difference scheme has been used to represent the term $\frac{\partial^2 u(x,t)}{\partial x^2}$ and a forward difference scheme for the derivative term $\frac{\partial u(x,t)}{\partial t}$. By substituting the differencing schemes of the derivative terms in Eq. (5.1), one may obtain

$$\frac{u_{i,j+1} - u_{i,j}}{\Delta t} = k \frac{u_{i+1,j} - 2u_{i,j} + u_{i-1,j}}{\Delta x^2} \qquad (5.2)$$

$$\Rightarrow u_{i,j+1} = r u_{i-1,j} + (1 - 2r) u_{i,j} + r u_{i+1,j}. \qquad (5.3)$$

where $r = \frac{\Delta t}{(\Delta x)^2}$.

Since, we know the values of $u_{i,j}$, $u_{i+1,j}$, and $u_{i-1,j}$ of Eq. (5.3), one may get the values of $u_{i,j+1}$.

5.3.2 Implicit Finite Difference Method

In implicit finite difference scheme the solution of governing differential equation depends on both the current state of the system and later state. So, a general recursive computation is not specified. Implicit schemes are generally solved by using matrix inverse methods for linear problems and iterative methods in nonlinear problems [10, 11].

Next, solving procedure has been included for a time-dependent diffusion equation by using the implicit finite difference method.

Let us consider a time-dependent equation

$$\frac{\partial u(x, t)}{\partial t} = k \frac{\partial^2 u(x, t)}{\partial x^2}, \qquad 0 \le x \le L, \quad t \ge 0 \qquad (5.4)$$

with the initial conditions

$$u(x, 0) = u_0(x) \Rightarrow u_{i,0} = u_0.$$

and the boundary conditions

$$u(0, t) = u_1(t) \Rightarrow u_{0,j+1} = u_1,$$
$$u(L, t) = u_2(t) \Rightarrow u_{N,j+1} = u_2.$$

By approximating the derivatives, Eq. (5.4) can be represented as

$$\frac{u_{i,j} - u_{i,j-1}}{\Delta t} = k \frac{u_{i+1,j} - 2u_{i,j} + u_{i-1,j}}{\Delta x^2} \qquad (5.5)$$

$$\Rightarrow u_{i,j-1} = -r u_{i+1,j} + (1 + 2r) u_{i,j} - r u_{i-1,j}. \qquad (5.6)$$

where $r = \frac{k \Delta t}{\Delta x^2}$.

There are different stability conditions for the above two schemes and interested readers may find the details in Refs. [1, 2, 4, 6, 10, 11].

The next section implements numerical example results of the explicit and implicit schemes for heat equation subject to specific initial and boundary conditions.

Example 5.1 Solve the heat equation by using explicit and implicit finite difference schemes

$$\frac{\partial u(x,t)}{\partial t} = \frac{\partial^2 u(x,t)}{\partial x^2} \tag{5.7}$$

with the initial condition $u(x,0) = 3\sin(\pi x)$ and boundary conditions $u(0,t) = 0$, $u(1,t) = 0$.

Solution By comparing the given problem (5.7) with Eq. (5.1) gives values as $k = 1$, $u_0(x) = 3\sin(\pi x)$, $u_1(x) = 0$, $u_2(x) = 0$, and $L = 1$.

Solving by Explicit Finite Difference Scheme By applying explicit finite difference representation at the general grid points (i, j) to the given Eq. (5.7) $\left(\frac{\partial u(x,t)}{\partial t}\right)_{i,j} = \left(\frac{\partial^2 u(x,t)}{\partial x^2}\right)_{i,j}$ with step lengths as $\Delta t = 0.001$ and $\Delta x = 0.05$, one may be obtain the difference scheme as follows from Eq. (5.3):

$$u_{i,j+1} = (0.4)u_{i-1,j} + (0.2)u_{i,j} + (0.4)u_{i+1,j}, \tag{5.8}$$

where $r = \frac{k\Delta t}{\Delta x^2} = 0.4$.

By finding the iteration values from Eq. (5.8), Figure 5.2 depicts the solution of Eq. (5.7) by using the explicit scheme.

Solving by Implicit Finite Difference Scheme In similar fashion, by applying implicit finite difference representation to the given Eq. (5.7) with step lengths as $\Delta t = 0.001$ and $\Delta x = 0.05$, one may be obtain the difference scheme as follows from Eq. (5.6):

$$u_{i,j-1} = -(0.4)u_{i+1,j} + (1.8)u_{i,j} - (0.4)u_{i-1,j} \tag{5.9}$$

where $r = \frac{k\Delta t}{\Delta x^2} = 0.4$.

By iterating Eq. (5.9), Figure 5.3 depicts the solution of Eq. (5.7) by using the implicit scheme.

In the next example we present a common elliptic PDE to determine the temperature at the inside nodes.

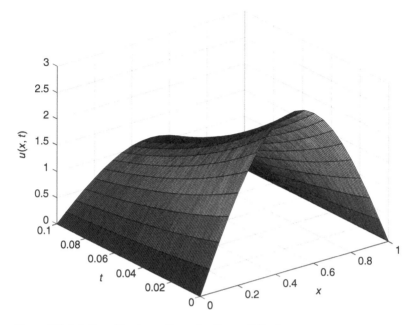

Figure 5.2 Solution of heat equation using the explicit scheme.

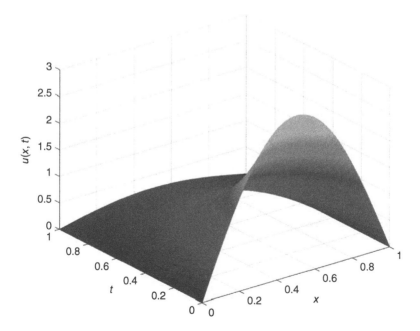

Figure 5.3 Solution of heat equation using the implicit scheme.

Example 5.2 Determine the temperature (T) at the inside nodes of the following square-shaped Figure 5.4, made up of nine equally sized squares, whereby the temperature along the boundaries are known.

Solution This is a boundary value problem for the Laplace equation:

$$\frac{\partial^2 T}{\partial x^2} + \frac{\partial^2 T}{\partial y^2} = 0 \qquad (5.10)$$

Apply the central differencing scheme to approximate the PDE (5.10) as

$$T_{i-1,j} + T_{i,j+1} - 4T_{i,j} + T_{i+1,j} + T_{i,j-1} = 0 \qquad (5.11)$$

Using Eq. (5.11) for each nodes 1–4, we get the system of equations as

Node 1: At this node from Figure 5.4, $T_{i,j} = T_1$, $T_{i-1,j} = 10$, $T_{i,j-1} = T_2$, $T_{i,j+1} = 20$, and $T_{i+1,j} = T_4$. By substituting in Eq. (5.11)

$$-4T_1 + T_2 + T_4 = -30 \qquad (5.12)$$

Node 2: In similar fashion $T_{i,j} = T_2$, $T_{i-1,j} = 10$, $T_{i,j-1} = 0$, $T_{i,j+1} = T_1$, and $T_{i+1,j} = T_3$. By replacing in Eq. (5.11)

$$T_1 - 4T_2 + T_3 = -10 \qquad (5.13)$$

Node 3: $T_{i,j} = T_3$, $T_{i-1,j} = T_2$, $T_{i,j-1} = 0$, $T_{i,j+1} = T_4$, and $T_{i+1,j} = 30$. From Eq. (5.11) we get

$$T_2 - 4T_3 + T_4 = -30 \qquad (5.14)$$

Node 4: From Figure 5.4, $T_{i,j} = T_4$, $T_{i-1,j} = T_1$, $T_{i,j-1} = T_3$, $T_{i,j+1} = 20$, and $T_{i+1,j} = 30$. By substituting in Eq. (5.11)

$$T_1 + T_3 - 4T_4 = -50 \qquad (5.15)$$

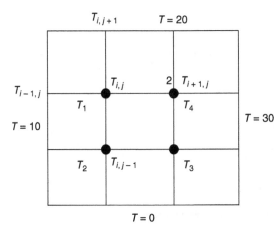

Figure 5.4 Discretizing the temperature (T) of a square shaped domain.

The four Eqs. (5.12)–(5.15) with four unknowns can also be written in the matrix form as

$$
\begin{bmatrix}
-4 & 1 & 0 & 1 \\
1 & -4 & 1 & 0 \\
0 & 1 & -4 & 1 \\
1 & 0 & 1 & -4
\end{bmatrix}
\begin{bmatrix}
T_1 \\
T_2 \\
T_3 \\
T_4
\end{bmatrix}
=
\begin{bmatrix}
-30 \\
-10 \\
-30 \\
-50
\end{bmatrix}
\tag{5.16}
$$

By solving the above system (5.16), one can determine the temperature at the inside nodes of Figure 5.4 as $T_1 = 15$, $T_2 = 10$, $T_3 = 15$, and $T_4 = 20$.

For more details of the FDMs, interested readers may refer Refs. [1–4, 6, 7, 9–11].

Exercise

1 Use FDMs to solve the boundary value problem $\frac{d^2y}{dx^2} + y = 1$ with respect to the boundary conditions $y(0) = 1$ and $y\left(\frac{\pi}{2}\right) = 0$.

2 Solve $y'' + \frac{1}{x}y' + \left(1 - \frac{1}{4x^2}\right)y = x$ with $y(1) = 1.0$ and $y(6) = -0.5$ using the FDM with 25 subintervals.

References

1 Hoffman, J.D. and Frankel, S. (2001). *Numerical Methods for Engineers and Scientists*. New York: CRC Press.

2 Bhat, R.B. and Chakraverty, S. (2004). *Numerical Analysis in Engineering*. London: Alpha Science International Ltd.

3 Euler, L. (2012). *Elements of Algebra*. New York: Springer Science & Business Media.

4 Iserles, A. (2009). *A First Course in the Numerical Analysis of Differential Equations* (No. 44). Cambridge, New York: Cambridge University Press.

5 Steger, J.L. (1978). Implicit finite-difference simulation of flow about arbitrary two-dimensional geometries. *AiAA Journal* 16 (7): 679–686.

6 Gel'fond, A.O. (1971). *Calculus of Finite Differences*. New Delhi: Hindustan Publishing Corporation.

7 Lele, S.K. (1992). Compact finite difference schemes with spectral-like resolution. *Journal of Computational Physics* 103 (1): 16–42.

8 Smith, G.D. (1985). *Numerical Solution of Partial Differential Equations: Finite Difference Methods*. Oxford: Oxford University Press.

9 Grossmann, C., Roos, H.G., and Stynes, M. (2007). *Numerical Treatment of Partial Differential Equations*, vol. 154. Berlin: Springer.

10 Ghazizadeh, H.R., Maerefat, M., and Azimi, A. (2010). Explicit and implicit finite difference schemes for fractional Cattaneo equation. *Journal of Computational Physics* 229 (19): 7042–7057.

11 Hughes, T.J. and Liu, W.K. (1978). Implicit-explicit finite elements in transient analysis: stability theory. *Journal of Applied Mechanics* 45 (2): 371–374.

6

Finite Element Method

6.1 Introduction

Finite element method (FEM) has wide applications in various science and engineering fields viz. structural mechanics, biomechanics and electromagnetic field problems, etc. of which exact solutions may not be determined. The FEM serves as a numerical discretization approach that converts differential equations into algebraic equations. In this regard, the weighted residual methods discussed in Chapter 3 convert governing differential equation to a system of algebraic equations over the entire domain Ω, whereas FEM is applied to finitely discretized elements of the global domain Ω. Also, the finite difference technique discussed in Chapter 5 generally considers the spacing of nodes such that the entire domain is partitioned in terms of squares and rectangles. The FEM overcomes this drawback as depicted in Figure 6.1 by spacing the nodes such that the entire domain is partitioned using any shape in general.

In this chapter simple differential equations are solved in order to have better understanding of the finite element technique. There exist various standard books [1–7] and the references mentioned therein devoted to the FEM and effective applications in various science and engineering fields. As such, this section gives an introduction to the FEM. Then, procedures of FEM and Galerkin FEM are discussed in Sections 6.2 and 6.3, respectively. Finally, the last section implements FEM for solving one-dimensional structural problems.

6.2 Finite Element Procedure

In the FEM, the global domain Ω is partitioned finitely to a number of nonoverlapping subdomains known as *finite elements*. Generally, a two-dimensional (2D) domain is partitioned in terms of 2D geometrical regions viz. triangles and quadrilaterals whereas a three-dimensional (3D) domain is partitioned using

Advanced Numerical and Semi-Analytical Methods for Differential Equations, First Edition.
Snehashish Chakraverty, Nisha Rani Mahato, Perumandla Karunakar, and Tharasi Dilleswar Rao.
© 2019 John Wiley & Sons, Inc. Published 2019 by John Wiley & Sons, Inc.

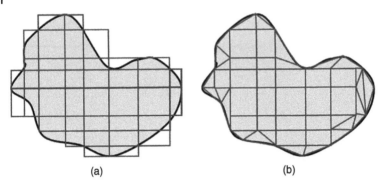

Figure 6.1 (a) Finite difference and (b) finite element discretization of arbitrary object.

3D regions such as cuboids, tetrahedrons, pyramids, etc. Finite element partition of complex structural systems using different shape elements may be found in Refs. [8–10]. In this regard, the procedures involved in the FEM [8] are formulated as below:

1. Divide the global domain $\Omega = [a, b]$ into finitely n subdomains or elements (Figure 6.2) as $\Omega_1^{(e)}, \Omega_2^{(e)}, \ldots, \Omega_n^{(e)}$ such that $\Omega_i^{(e)} = [x_{i-1}, x_i]$ for $i = 1, 2, \ldots, n$. Recall that x_i are referred to as $n+1$ nodes or nodal points where x_i are not necessarily to be evenly spaced ($\Delta x_i = x_i - x_{i-1}$ is not same for $i = 1, 2, \ldots, n$).

2. The approximate solution $\hat{u}(x)$ for the differential equation,

$$L(u) + N(u) = f(x) \tag{6.1}$$

subject to boundary conditions $u(a) = \gamma_a$ and $u(b) = \gamma_b$ is expressed in terms of element interpolating polynomials $u_i^{(e)}(x)$,

$$\hat{u}(x) = \sum_{i=1}^{n} u_i^{(e)}(x). \tag{6.2}$$

3. Consider the element interpolating functions $u_i^{(e)}(x)$ as

$$u_i^{(e)}(x) = \phi_{i-1}^{(i)}(x)u_{i-1} + \phi_i^{(i)}(x)u_i. \tag{6.3}$$

Here, ϕ are the interpolating functions that resemble the *shape functions*. For instance, the first-degree Lagrange polynomial (given in Figure 6.3) for

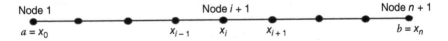

Figure 6.2 Nodal points for the domain $\Omega = [a, b]$.

Figure 6.3 Shape function of Lagrange polynomial.

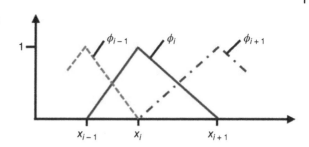

each element is

$$u_i^{(e)}(x) = \frac{x - x_i}{x_{i-1} - x_i} u_{i-1} + \frac{x - x_{i-1}}{x_i - x_{i-1}} u_i$$

$$\Rightarrow u_i^{(e)}(x) = -\frac{x - x_i}{\Delta x_i} u_{i-1} + \frac{x - x_{i-1}}{\Delta x_i} u_i \tag{6.4}$$

where the interval spacing is given by $\Delta x_i = x_i - x_{i-1}$.

Other interpolation polynomial approximations viz. Hermite [9, 10] may also be used.

4. Obtain the system of equations known as *element equations* involving nodal values for each element that may be found using the Galerkin-weighted residual method discussed in Chapter 3.

5. Combine the set of element equations obtained using the Galerkin method. This method of combination is generally referred to as *assembling*. Apply the Galerkin-weighted residual method to each element between the nodes x_{i-1} and x_i to obtain the *nodal equations*.

6. Solve the assembled system of equations involving nodal points to obtain global approximate solution of governing differential equation for the entire domain.

For ease of understanding the FEM, the step-by-step procedure of linear second-order ordinary differential equation (ODE) is considered with respect to the Galerkin method in Section 6.3.1.

6.3 Galerkin Finite Element Method

In this section, the steps [8, 9] involved in solving differential equations [11, 12] using the above-mentioned procedure is discussed.

6.3.1 Ordinary Differential Equation

This section gives a detailed procedure for solving ODE subject to boundary conditions using the Galerkin FEM.

Let us consider a linear second-order ODE,

$$\frac{d^2u}{dx^2} + Q(x)u = f(x), \quad u(a) = \gamma_a, \quad u(b) = \gamma_b \tag{6.5}$$

over the domain $[a, b]$.

Step (i): Divide $[a, b]$ into finitely n elements $\Omega_1^{(e)} = [x_0, x_1] = [a, x_1]$, $\Omega_2^{(e)} = [x_1, x_2], \ldots, \Omega_n^{(e)} = [x_{n-1}, x_n] = [x_{n-1}, b]$ such that $x_0, x_1, \ldots, x_n$ are nodes.

Step (ii): Using Eq. (6.4), the element interpolating polynomial is considered as

$$u_i^{(e)}(x) = -\frac{x - x_i}{\Delta x_i}u_{i-1} + \frac{x - x_{i-1}}{\Delta x_i}u_i, \quad \text{for} \quad i = 1, 2, \ldots, n$$

having element shape functions

$$\phi_{i-1}^{(i)} = -\frac{x - x_i}{\Delta x_i} \quad \text{and} \quad \phi_i^{(i)} = \frac{x - x_{i-1}}{\Delta x_i}. \tag{6.6}$$

Step (iii): Apply the Galerkin method to each element i with respect to the weight function $\phi_k(x)$, where $k = i - 1, i$ for obtaining element equations.

The weighted residual integrals $I(u(x))$ are obtained as

$$I(u(x)) = \int_{x_{i-1}}^{x_i} \phi_{i-1}^{(i)}(x)R(x)dx = 0 \Rightarrow \int_{x_{i-1}}^{x_i} \phi_{i-1}^{(i)}(x)\left(\frac{d^2u}{dx^2} + Q(x)u - f(x)\right) dx = 0 \tag{6.7a}$$

and

$$I(u(x)) = \int_{x_{i-1}}^{x_i} \phi_i^{(i)}(x)R(x)dx = 0 \Rightarrow \int_{x_{i-1}}^{x_i} \phi_i^{(i)}(x)\left(\frac{d^2u}{dx^2} + Q(x)u - f(x)\right) dx = 0 \tag{6.7b}$$

where $R(x) = \frac{d^2u}{dx^2} + Q(x)u - f(x)$ is the residual. Equations (6.7a) and (6.7b) further get reduced to

$$\int_{x_{i-1}}^{x_i} \phi_{i-1}^{(i)}(x)\frac{d^2u}{dx^2}dx + \int_{x_{i-1}}^{x_i} \phi_{i-1}^{(i)}(x)Q(x)udx - \int_{x_{i-1}}^{x_i} \phi_{i-1}^{(i)}(x)f(x)dx = 0 \tag{6.8a}$$

and

$$\int_{x_{i-1}}^{x_i} \phi_i^{(i)}(x)\frac{d^2u}{dx^2}dx + \int_{x_{i-1}}^{x_i} \phi_i^{(i)}(x)Q(x)udx - \int_{x_{i-1}}^{x_i} \phi_i^{(i)}(x)f(x)dx = 0. \tag{6.8b}$$

By applying integration by parts to Eqs. (6.8a) and (6.8b), we get

$$-\int_{x_{i-1}}^{x_i} \left(\frac{d\phi_{i-1}^{(i)}}{dx}\right)\left(\frac{du}{dx}\right) dx + \int_{x_{i-1}}^{x_i} \phi_{i-1}^{(i)}Qudx - \int_{x_{i-1}}^{x_i} \phi_{i-1}^{(i)}fdx$$

$$+ \phi_{i-1}^{(i)}\frac{du}{dx}\bigg|_{x=x_i} - \phi_{i-1}^{(i)}\frac{du}{dx}\bigg|_{x=x_{i-1}} = 0 \tag{6.9a}$$

and

$$-\int_{x_{i-1}}^{x_i} \left(\frac{d\phi_i^{(i)}}{dx}\right)\left(\frac{du}{dx}\right)dx + \int_{x_{i-1}}^{x_i} \phi_i^{(i)} Q u dx - \int_{x_{i-1}}^{x_i} \phi_i^{(i)} f dx$$

$$+ \phi_i^{(i)}\frac{du}{dx}\Big|_{x=x_i} - \phi_i^{(i)}\frac{du}{dx}\Big|_{x=x_{i-1}} = 0. \tag{6.9b}$$

The shape function in the Galerkin method is selected such that $\phi_{i-1}^{(i)}$ is zero in all integrals except at ith integral and $\phi_{i-1}^{(i)}(a) = \phi_i^{(i)}(b) = 0$. As such, Eqs. (6.9a) and (6.9b) further reduce to

$$-\int_{x_{i-1}}^{x_i} \left(\frac{d\phi_{i-1}^{(i)}}{dx}\right)\left(\frac{du}{dx}\right)dx + \int_{x_{i-1}}^{x_i} \phi_{i-1}^{(i)} Q u dx - \int_{x_{i-1}}^{x_i} \phi_{i-1}^{(i)} f dx = 0 \tag{6.10a}$$

and

$$-\int_{x_{i-1}}^{x_i} \left(\frac{d\phi_i^{(i)}}{dx}\right)\left(\frac{du}{dx}\right)dx + \int_{x_{i-1}}^{x_i} \phi_i^{(i)} Q u dx - \int_{x_{i-1}}^{x_i} \phi_i^{(i)} f dx = 0. \tag{6.10b}$$

By using the shape functions given in Eq. (6.6), we have

$$\frac{d\phi_{i-1}^{(i)}}{dx} = -\frac{1}{\Delta x_i} \quad \text{and} \quad \frac{d\phi_i^{(i)}}{dx} = \frac{1}{\Delta x_i}. \tag{6.11}$$

By substituting Eq. (6.11) in Eqs. (6.10a) and (6.10b), one may obtain

$$\frac{1}{\Delta x_i}\int_{x_{i-1}}^{x_i} \left(\frac{du}{dx}\right)dx + \int_{x_{i-1}}^{x_i} \phi_{i-1}^{(i)} Q u dx - \int_{x_{i-1}}^{x_i} \phi_{i-1}^{(i)} f dx = 0 \tag{6.12a}$$

and

$$-\frac{1}{\Delta x_i}\int_{x_{i-1}}^{x_i} \left(\frac{du}{dx}\right)dx + \int_{x_{i-1}}^{x_i} \phi_i^{(i)} Q u dx - \int_{x_{i-1}}^{x_i} \phi_i^{(i)} f dx = 0. \tag{6.12b}$$

Taking average values for Q and f over the integral, Eqs. (6.12a) and (6.12b) reduce to

$$\frac{1}{\Delta x_i}\int_{x_{i-1}}^{x_i} \left(\frac{du}{dx}\right)dx + Q_{av}^{(i)}\int_{x_{i-1}}^{x_i} \phi_{i-1}^{(i)} u dx - f_{av}^{(i)}\int_{x_{i-1}}^{x_i} \phi_{i-1}^{(i)} dx = 0 \tag{6.13a}$$

and

$$-\frac{1}{\Delta x_i}\int_{x_{i-1}}^{x_i} \left(\frac{du}{dx}\right)dx + Q_{av}^{(i)}\int_{x_{i-1}}^{x_i} \phi_i^{(i)} u dx - f_{av}^{(i)}\int_{x_{i-1}}^{x_i} \phi_i^{(i)} dx = 0 \tag{6.13b}$$

where $Q_{av}^{(i)} = \frac{Q_{i-1}+Q_i}{2}$ and $f_{av}^{(i)} = \frac{f_{i-1}+f_i}{2}$. Moreover, $\frac{du}{dx} = \frac{d}{dx}u_i^{(e)}(x) = -\frac{u_{i-1}}{\Delta x_i} + \frac{u_i}{\Delta x_i}$, which yields

$$\frac{du}{dx} = \frac{u_i - u_{i-1}}{\Delta x_i}. \tag{6.14}$$

Equations (6.13a) and (6.13b) reduce to system of equations (element equations) for ith element,

$$-\left(\frac{1}{\Delta x_i} - \frac{Q_{av}^{(i)}\Delta x_i}{3}\right)u_{i-1} + \left(\frac{1}{\Delta x_i} + \frac{Q_{av}^{(i)}\Delta x_i}{6}\right)u_i - \frac{f_{av}^{(i)}\Delta x_i}{2} = 0 \quad (6.15a)$$

and

$$\left(\frac{1}{\Delta x_i} + \frac{Q_{av}^{(i)}\Delta x_i}{6}\right)u_{i-1} - \left(\frac{1}{\Delta x_i} - \frac{Q_{av}^{(i)}\Delta x_i}{3}\right)u_i - \frac{f_{av}^{(i)}\Delta x_i}{2} = 0. \quad (6.15b)$$

In case of equal spacing, the ith element system of equations gets transformed to

$$-\left(\frac{1}{\Delta x} - \frac{Q_{av}^{(i)}\Delta x}{3}\right)u_{i-1} + \left(\frac{1}{\Delta x} + \frac{Q_{av}^{(i)}\Delta x}{6}\right)u_i - \frac{f_{av}^{(i)}\Delta x}{2} = 0 \quad (6.16a)$$

and

$$\left(\frac{1}{\Delta x} + \frac{Q_{av}^{(i)}\Delta x}{6}\right)u_{i-1} - \left(\frac{1}{\Delta x} - \frac{Q_{av}^{(i)}\Delta x}{3}\right)u_i - \frac{f_{av}^{(i)}\Delta x}{2} = 0. \quad (6.16b)$$

Step (iv): The element equations are then assembled to form a system of equations with unknowns $u_0, u_1,\ldots, u_n$.

By using Eqs. (6.16a) and (6.16b), the element equations for $(i+1)$th element is given by

$$-\left(\frac{1}{\Delta x_{i+1}} - \frac{Q_{av}^{(i+1)}\Delta x_{i+1}}{3}\right)u_i + \left(\frac{1}{\Delta x_{i+1}} + \frac{Q_{av}^{(i+1)}\Delta x_{i+1}}{6}\right)u_{i+1} - \frac{f_{av}^{(i+1)}\Delta x_{i+1}}{2} = 0$$

$$(6.17a)$$

and

$$\left(\frac{1}{\Delta x_{i+1}} + \frac{Q_{av}^{(i+1)}\Delta x_{i+1}}{6}\right)u_i - \left(\frac{1}{\Delta x_{i+1}} - \frac{Q_{av}^{(i+1)}\Delta x_{i+1}}{3}\right)u_{i+1} - \frac{f_{av}^{(i+1)}\Delta x_{i+1}}{2} = 0.$$

$$(6.17b)$$

Hence, the nodal equation at $(i+1)$th node is obtained by adding Eqs. (6.15b) and (6.17a),

$$\left(\frac{1}{\Delta x_i} + \frac{Q_{av}^{(i)}\Delta x_i}{6}\right)u_{i-1} - \left(\frac{1}{\Delta x_i} + \frac{1}{\Delta x_{i+1}} - \frac{Q_{av}^{(i)}\Delta x_i + Q_{av}^{(i+1)}\Delta x_{i+1}}{6}\right)u_i$$

$$+\left(\frac{1}{\Delta x_{i+1}} + \frac{Q_{av}^{(i+1)}\Delta x_{i+1}}{6}\right)u_{i+1} = \frac{f_{av}^{(i)}\Delta x_i + f_{av}^{(i+1)}\Delta x_{i+1}}{2}. \quad (6.18)$$

Then, Eq. (6.18) for equal spacing yields

$$\left(\frac{1}{\Delta x} + \frac{Q_{av}^{(i)}\Delta x}{6}\right)u_{i-1} - 2\left(\frac{1}{\Delta x} - \frac{(Q_{av}^{(i)} + Q_{av}^{(i+1)})\Delta x}{6}\right)u_i$$

$$+ \left(\frac{1}{\Delta x} + \frac{Q_{av}^{(i+1)}\Delta x}{6}\right)u_{i+1} = \frac{(f_{av}^{(i)} + f_{av}^{(i+1)})\Delta x}{2} \tag{6.19}$$

Step (v): Adjust the obtained system of equations using the boundary conditions $u_0 = u(a) = \gamma_a$ and $u_n = u(b) = \gamma_b$ for unknowns $u_1, u_2, \ldots, u_{n-1}$.

Step (vi): Solve the adjusted system for intermediate nodal values of $u(x)$, that is $u_1, u_2, \ldots, u_{n-1}$.

In order to have a better understanding, let us now consider an example for solving linear ODE.

Example 6.1 Solve the boundary value problem (Example 3.4),

$$u'' = x \tag{6.20}$$

subject to $u(0) = 1$ and $u(2) = 2$ using the Galerkin FEM.

Solution Let us consider the partitioning of $\Omega = [0, 2]$ into eight elements as depicted in Figure 6.4, $\Omega_1^{(e)} = \left[0, \frac{1}{4}\right]$, $\Omega_2^{(e)} = \left[\frac{1}{4}, \frac{1}{2}\right]$, $\Omega_3^{(e)} = \left[\frac{1}{2}, \frac{3}{4}\right]$, $\Omega_4^{(e)} = \left[\frac{3}{4}, 1\right]$, $\Omega_5^{(e)} = \left[1, \frac{5}{4}\right]$, $\Omega_6^{(e)} = \left[\frac{5}{4}, \frac{3}{2}\right]$, $\Omega_7^{(e)} = \left[\frac{3}{2}, \frac{7}{4}\right]$, and $\Omega_8^{(e)} = \left[\frac{7}{4}, 2\right]$.

The above partitions are equally spaced. So, we obtain the nodal equations at interior nodes 2–8 by plugging $i = 1$–7, respectively, in Eq. (6.19), where $\Delta x = \frac{1}{4}$, $Q = 0, f(x) = x, a = 0$, and $b = 2$.

Node 2: $\qquad 4u_0 - 8u_1 + 4u_2 = \dfrac{1}{16}$

Node 3: $\qquad 4u_1 - 8u_2 + 4u_3 = \dfrac{1}{8}$

Node 4: $\qquad 4u_2 - 8u_3 + 4u_4 = \dfrac{3}{16}$

Node 5: $\qquad 4u_3 - 8u_4 + 4u_5 = \dfrac{1}{4}$

Node 6: $\qquad 4u_4 - 8u_5 + 4u_6 = \dfrac{5}{16}$

Node 7: $\qquad 4u_5 - 8u_6 + 4u_7 = \dfrac{3}{8}$

Node 8: $\qquad 4u_6 - 8u_7 + 4u_8 = \dfrac{7}{16}$

Node 1  Node 9

$x_0 = 0$ 1 $\frac{1}{4}$ $\frac{1}{2}$ $\frac{3}{4}$ 1 $\frac{5}{4}$ $\frac{3}{2}$ $\frac{7}{4}$ $x_8 = 2$

Figure 6.4 Nodal points for the domain $\Omega = [0, 2]$.

The nodal equations at end nodes 1 and 9 are obtained by plugging $i = 1$ and 7 in Eqs. (6.15a) and (6.17b), respectively,

Node 1: $\qquad -4u_0 + 4u_1 = \dfrac{1}{64}$

Node 9: $\qquad 4u_7 - 4u_8 = \dfrac{15}{64}$

Accordingly, the system of equations is obtained as

$$
\begin{bmatrix}
4 & -4 \\
-4 & 8 & -4 \\
 & -4 & 8 & -4 \\
 & & -4 & 8 & -4 \\
 & & & -4 & 8 & -4 \\
 & & & & -4 & 8 & -4 \\
 & & & & & -4 & 8 & -4 \\
 & & & & & & -4 & 8 & -4 \\
 & & & & & & & -4 & 4
\end{bmatrix}
\begin{Bmatrix}
u_0 \\ u_1 \\ u_2 \\ u_3 \\ u_4 \\ u_5 \\ u_6 \\ u_7 \\ u_8
\end{Bmatrix}
=
\begin{Bmatrix}
-0.0156 \\ -0.0625 \\ -0.125 \\ -0.1875 \\ -0.25 \\ -0.3125 \\ -0.375 \\ -0.4375 \\ -0.2344
\end{Bmatrix}
$$

$$(6.21)$$

By using boundary conditions $\gamma_a = 1$ and $\gamma_b = 2$, Eq. (6.21) is reduced to

$$
\begin{bmatrix}
8 & -4 \\
-4 & 8 & -4 \\
 & -4 & 8 & -4 \\
 & & -4 & 8 & -4 \\
 & & & -4 & 8 & -4 \\
 & & & & -4 & 8 & -4 \\
 & & & & & -4 & 8
\end{bmatrix}
\begin{Bmatrix}
u_1 \\ u_2 \\ u_3 \\ u_4 \\ u_5 \\ u_6 \\ u_7
\end{Bmatrix}
=
\begin{Bmatrix}
3.9375 \\ -0.125 \\ -0.1875 \\ -0.25 \\ -0.3125 \\ -0.375 \\ 7.5625
\end{Bmatrix}
\qquad (6.22)
$$

resulting in the intermediate nodal values $u_1 = 0.9609$, $u_2 = 0.9375$, $u_3 = 0.9453$, $u_4 = 1$, $u_5 = 1.1172$, $u_6 = 1.3125$, and $u_7 = 1.6016$. Then, the obtained nodal values are compared with exact solution $u(x) = 1 - \frac{x}{6} + \frac{x^3}{6}$ in Figure 6.5.

Figure 6.5 Comparison of the Galerkin FEM solution with the exact solution for Example 6.1.

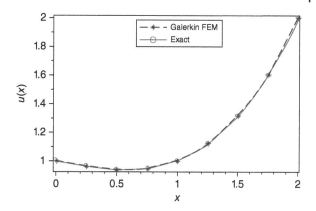

6.3.2 Partial Differential Equation

In this section, the Galerkin FEM is applied to solve diffusion equation

$$\frac{\partial u}{\partial t} = \alpha\frac{\partial^2 u}{\partial x^2} + Q(x)u - f(x) \tag{6.23}$$

satisfying boundary conditions $u(a, t) = \gamma_a$ and $u(b, t) = \gamma_b$ over the domain $\Omega(x, t)$.

The steps involved in solving diffusion equation using the Galerkin FEM are illustrated below:

Step (i): Divide the global domain $\Omega(x, t)$ into finitely $m \times n$ elements as illustrated in Figure 6.6, with respect to time step $\Delta t_j = t_j - t_{j-1}$ and element length step $\Delta x_i = x_i - x_{i-1}$. In case of equal spacing discretization, $\Delta t_j = \Delta t$ and $\Delta x_i = \Delta x$ for $j = 1, 2, \ldots, n$ and $i = 1, 2, \ldots, m$.

Figure 6.6 Finite-element two-dimensional discretization.

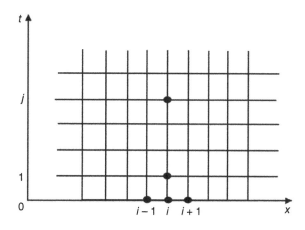

Step (ii): Consider the element interpolating polynomial for the global approximate solution $\hat{u}(x, t) = \sum_{i=1}^{m} u_i^j(x, t)$ as

$$u_i^j(x, t) = u_i^{(e)}(x, t) = -\frac{x - x_i}{\Delta x_i} u_{i-1}(t) + \frac{x - x_{i-1}}{\Delta x_i} u_i(t), \tag{6.24}$$

for $i = 1, 2, \ldots, m$ having element shape functions

$$\phi_{i-1}(x) = -\frac{x - x_i}{\Delta x_i} \quad \text{and} \quad \phi_i(x) = \frac{x - x_{i-1}}{\Delta x_i}.$$

Step (iii): Apply the Galerkin FEM [9] to each element i in order to obtain the corresponding element equations.

In this regard, the weighted residual integral is given as

$$I(u(x, t)) = \int_a^b w \left(\frac{\partial u}{\partial t} - \alpha \frac{\partial^2 u}{\partial x^2} - Q(x)u + f(x) \right) dx = 0 \tag{6.25}$$

where w is the weight function and $R(t, x) = \frac{\partial u}{\partial t} - \alpha \frac{\partial^2 u}{\partial x^2} - Q(x)u + f(x)$ is the residual.

Equation (6.25) is integrated by parts such that $-\int_a^b w\alpha \frac{\partial^2 u}{\partial x^2} dx = \int_a^b \alpha \frac{dw}{dx} \frac{du}{dx} dx - w\alpha \frac{\partial u}{\partial x} \Big|_a^b$. Here, $w\alpha \frac{\partial u}{\partial x} \Big|_a^b$ cancels out in all the interior nodes (except at end nodes) during assembling. As such, Eq. (6.25) reduces to

$$I(u(x, t)) = \int_a^b \left(w \frac{\partial u}{\partial t} + \alpha \frac{dw}{dx} \frac{du}{dx} - wQu + wf \right) dx = 0. \tag{6.26}$$

For each ith element, Eq. (6.26) reduces to element integral,

$$I_i(u_i(x, t)) = \int_{x_{i-1}}^{x_i} \left(w^{(i)} \frac{\partial u_i^{(e)}}{\partial t} + \alpha \frac{dw^{(i)}}{dx} \frac{\partial u_i^{(e)}}{\partial x} - w^{(i)} Q u_i^{(e)} + w^{(i)} f \right) dx = 0. \tag{6.27}$$

Here, Eq. (6.27) involves partial derivatives

$$\frac{\partial u_i^{(e)}}{\partial t} = \frac{\partial u_{i-1}}{\partial t} \left(-\frac{x - x_i}{\Delta x_i} \right) + \frac{\partial u_i}{\partial t} \left(\frac{x - x_{i-1}}{\Delta x_i} \right) \quad \text{and} \quad \frac{\partial u_i^{(e)}}{\partial x} = \frac{u_i - u_{i-1}}{\Delta x_i}.$$

Moreover, the weight functions in the Galerkin-weighted residual method are considered in terms of the shape functions $\phi_{i-1}(x)$ and $\phi_i(x)$ as

$$w^{(i)}(x) = \phi_{i-1}(x) = -\frac{x - x_i}{\Delta x_i} \quad \text{and} \quad w^{(i)}(x) = \phi_i(x) = \frac{x - x_{i-1}}{\Delta x_i}$$

$$\Rightarrow \frac{dw^{(i)}}{dx} = -\frac{1}{\Delta x_i}.$$

Using the weight function $w^{(i)} = \phi_{i-1}$, Eq. (6.27) reduces to

$$I_i(u_i(x,t)) = \frac{\Delta x_i}{6}\left(2\frac{\partial u_{i-1}}{\partial t} + \frac{\partial u_i}{\partial t}\right) - \frac{\alpha}{\Delta x_i}(u_i - u_{i-1})$$

$$- \frac{\Delta x_i Q_{av}^{(i)}}{6}(2u_{i-1} + u_i) + \frac{\Delta x_i f_{av}^{(i)}}{2} = 0 \qquad (6.28a)$$

where $Q_{av}^{(i)} = \frac{Q_{i-1}+Q_i}{2}$ and $f_{av}^{(i)} = \frac{f_{i-1}+f_i}{2}$ are the average values of $Q(x)$ and $f(x)$, respectively, for ith element. Further, substituting $w^{(i)} = \phi_i$ in Eq. (6.27) results in

$$I_i(u_i(x,t)) = \frac{\Delta x_i}{6}\left(\frac{\partial u_{i-1}}{\partial t} + 2\frac{\partial u_i}{\partial t}\right) + \frac{\alpha}{\Delta x_i}(u_i - u_{i-1})$$

$$- \frac{\Delta x_i Q_{av}^{(i)}}{6}(u_{i-1} + 2u_i) + \frac{\Delta x_i f_{av}^{(i)}}{2} = 0 \qquad (6.28b)$$

Equations (6.28a) and (6.28b) are referred to as the element equations for the ith element.

Step (iv): Obtain the nodal equations by assembling the element equations at $(i+1)$th node to form a system of equations with respect to unknown nodal values.

Using Eqs. (6.28a) and (6.28b), the element equations for the $(i+1)$th element are

$$\frac{\Delta x_{i+1}}{6}\left(2\frac{\partial u_i}{\partial t} + \frac{\partial u_{i+1}}{\partial t}\right) - \frac{\alpha}{\Delta x_{i+1}}(u_{i+1} - u_i) - \frac{\Delta x_{i+1}Q_{av}^{(i+1)}}{6}(2u_i + u_{i+1})$$

$$+ \frac{\Delta x_{i+1}f_{av}^{(i+1)}}{2} = 0, \qquad (6.29a)$$

$$\frac{\Delta x_{i+1}}{6}\left(\frac{\partial u_i}{\partial t} + 2\frac{\partial u_{i+1}}{\partial t}\right) + \frac{\alpha}{\Delta x_{i+1}}(u_{i+1} - u_i) - \frac{\Delta x_{i+1}Q_{av}^{(i+1)}}{6}(u_i + 2u_{i+1})$$

$$+ \frac{\Delta x_{i+1}f_{av}^{(i+1)}}{2} = 0. \qquad (6.29b)$$

The nodal equation at the $(i+1)$th node is obtained by adding Eqs. (6.28b) and (6.29a),

$$\frac{\partial u_{i-1}}{\partial t} + 4\frac{\partial u_i}{\partial t} + \frac{\partial u_{i+1}}{\partial t} + \frac{6\alpha(u_i - u_{i-1})}{\Delta x_i^2} - \frac{6\alpha(u_{i+1} - u_i)}{\Delta x_{i+1}^2}$$

$$- Q_{av}^{(i)}(u_{i-1} + 2u_i) - Q_{av}^{(i+1)}(2u_i + u_{i+1}) + 3(f_{av}^{(i)} + f_{av}^{(i+1)}) = 0. \qquad (6.30)$$

Using finite difference approximation $\frac{\partial u^j}{\partial t} = \frac{u^{j+1}-u^j}{\Delta t}$ for $j = 1, 2, \ldots, n$, we further obtain

$$u_{i-1}^{j+1} + 4u_i^{j+1} + u_{i+1}^{j+1}$$

$$= u_{i-1}^j + 4u_i^j + u_{i+1}^j - \frac{6\alpha\Delta t(u_i^j - u_{i-1}^j)}{\Delta x_i^2} + \frac{6\alpha\Delta t(u_{i+1}^j - u_i^j)}{\Delta x_{i+1}^2}$$

$$+ Q_{av}^{(i)}\Delta t(u_{i-1}^j + 2u_i^j) + Q_{av}^{(i+1)}\Delta t(2u_i^j + u_{i+1}^j) - 3\Delta t(f_{av}^{(i)} + f_{av}^{(i+1)}) = 0.$$
$$(6.31)$$

Equation (6.31) for equal spacing yields

$$u_{i-1}^{j+1} + 4u_i^{j+1} + u_{i+1}^{j+1}$$

$$= u_{i-1}^j + 4u_i^j + u_{i+1}^j + \frac{6\alpha\Delta t}{\Delta x^2}(u_{i-1}^j - 2u_i^j + u_{i+1}^j)$$

$$+ Q_{av}^{(i)}\Delta t(u_{i-1}^j + 2u_i^j) + Q_{av}^{(i+1)}\Delta t(2u_i^j + u_{i+1}^j) - 3\Delta t(f_{av}^{(i)} + f_{av}^{(i+1)}) = 0.$$
$$(6.32)$$

Step (v): Adjust the obtained system of equations using the boundary conditions $u(a, t) = \gamma_a$ and $u(b, t) = \gamma_b$ to obtain unknown nodal values.

Example 6.2 Let us consider the one-dimensional heat equation (Example 5.1),

$$u_t - u_{xx} = 0 \qquad (6.33)$$

subject to initial condition $u(x, 0) = 3\sin(\pi x)$ and boundary conditions $u(0, t) = 0$, $u(1, t) = 0$ for $\Delta t = 0.02$ and $\Delta x = 0.25$.

Solution Comparing with diffusion equation (6.23), we have $\alpha = 1$, $Q = 0$, and $f = 0$. The nodal equations at the intermediate nodes 2–4 are obtained by substituting $i = 1$–3, respectively, in Eq. (6.32).

Node 2: $\quad u_0^{j+1} + 4u_1^{j+1} + u_2^{j+1} = u_0^j + 4u_1^j + u_2^j + 1.92(u_0^j - 2u_1^j + u_2^j)$

Node 3: $\quad u_1^{j+1} + 4u_2^{j+1} + u_3^{j+1} = u_1^j + 4u_2^j + u_3^j + 1.92(u_1^j - 2u_2^j + u_3^j)$

Node 4: $\quad u_2^{j+1} + 4u_3^{j+1} + u_4^{j+1} = u_2^j + 4u_3^j + u_4^j + 1.92(u_2^j - 2u_3^j + u_4^j)$

By using the boundary conditions $u_0 = 0$ and $u_4 = 0$, the nodal equations reduce to

Node 2: $\quad 4u_1^{j+1} + u_2^{j+1} = 0.16u_1^j + 2.92u_2^j$

Node 3: $\quad u_1^{j+1} + 4u_2^{j+1} + u_3^{j+1} = 2.92u_1^j + 0.16u_2^j + 2.92u_3^j$

Node 4: $\quad u_2^{j+1} + 4u_3^{j+1} = 2.92u_2^j + 0.16u_3^j$

Then, the system of equation is obtained as

$$
\begin{bmatrix} 4 & 1 & 0 \\ 1 & 4 & 1 \\ 0 & 1 & 4 \end{bmatrix} \begin{Bmatrix} u_1^{j+1} \\ u_2^{j+1} \\ u_3^{j+1} \end{Bmatrix} = \begin{bmatrix} 0.16 & 2.92 & 0 \\ 2.92 & 0.16 & 2.92 \\ 0 & 2.92 & 0.16 \end{bmatrix} \begin{Bmatrix} u_1^{j} \\ u_2^{j} \\ u_3^{j} \end{Bmatrix}.
$$

By using the given initial condition $u(x, 0) = 3\sin(\pi x)$, we get the initial approximations as $u_1^0 = 2.1213$, $u_2^0 = 3$, and $u_3^0 = 2.1213$. Then, the iterative solution of Eq. (6.33) is obtained and incorporated into Table 6.1.

Note: In case of variation problems, the Galerkin finite element method may be replaced with the Rayleigh–Ritz finite element method [8, 9].

The Galerkin FEM for solving linear differential equations, outlined in Sections 6.3.1 and 6.3.2 may be easily extended for nonlinear problems. In case of nonlinear differential equations, two procedures may be followed:

1. Linearize the governing nonlinear differential equation to obtain a linear differential equation. Then, reduce the obtained linear differential equation to a system of linear equations for computing the approximate solution using the FEM.
2. Apply the FEM directly to convert the governing nonlinear differential equation to a system of nonlinear equations. Then, solve the obtained system using Newton's method for the approximate solution.

FEMs for solving nonlinear boundary value problems have been discussed in various standard books. Readers interested in solving nonlinear differential equations using the FEM are encouraged to refer [9, 10].

Table 6.1 Solution of one-dimensional heat equation by the Galerkin FEM.

			$u(x, t)$		
$t \backslash x$	0	0.25	0.5	0.75	1
0	0	2.1213	3	2.1213	0
0.02	0	1.6807	2.3768	1.6807	0
0.04	0	1.3315	1.8831	1.3315	0
0.06	0	1.0549	1.4919	1.0549	0
0.08	0	0.8358	1.1820	0.8358	0
0.1	0	0.6622	0.9364	0.6622	0

6.4 Structural Analysis Using FEM

In structural mechanics, the approximate solution of governing partial differential equations for structures may be obtained using the FEM. Under static and dynamic conditions, the associated differential equations get transformed to simultaneous algebraic equations and eigenvalue problems, respectively. In this regard, static and dynamic analysis of structural systems are considered in Sections 6.4.1 and 6.4.2 respectively.

6.4.1 Static Analysis

This section considers the FEM modeling of one-dimensional structural system subject to static conditions. For instance, consider the simplest one-dimensional finite-element structure of spring or bar given in Figure 6.7.

A spring having stiffness k or bar having Young's modulus E of length L may be considered as an element having two nodes with forces f_1, f_2 applied on either nodes 1 and 2, respectively. The system of equations for one element spring as given in Refs. [2, 8] is obtained as

$$\begin{bmatrix} k & -k \\ -k & k \end{bmatrix} \begin{Bmatrix} u_1 \\ u_2 \end{Bmatrix} = \begin{Bmatrix} f_1 \\ f_2 \end{Bmatrix} \tag{6.34}$$

such that u_1 and u_2 are the nodal displacements. For one-element bar having stiffness $k = \frac{EA}{L}$, the system is considered as given in Eq. (6.34) subject to axial force $f_2 = -f_1 = \frac{EA}{L}(u_2 - u_1)$, where A is the cross-sectional area of the bar.

Similarly, for n element bar given in Figure 6.8, the system of equations is obtained as

$$\begin{bmatrix} k_1 & -k_1 \\ -k_1 & k_1 + k_2 & -k_2 \\ & \ddots & \ddots & \ddots \\ & & -k_n & k_{n-1} + k_n & -k_n \\ & & & -k_n & k_n \end{bmatrix} \begin{Bmatrix} u_1 \\ u_2 \\ \vdots \\ u_n \\ u_{n+1} \end{Bmatrix} = \begin{Bmatrix} f_1 \\ f_2 \\ \vdots \\ f_n \\ f_{n+1} \end{Bmatrix} \tag{6.35}$$

Node 1 —W— Node 2　　Node 1 E Node 2　　**Figure 6.7** One-element (a) spring and
k　　　　　　　　L　　　　(b) bar.
(a)　　　　　　　(b)

Node 1　Node 2　...　Node n　Node $n + 1$　**Figure 6.8** n bar element.

L

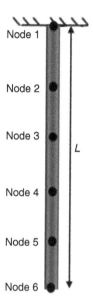

Figure 6.9 Five-element hanging rod.

having element length l_i, modulus of elasticity E_i, and area A_i such that stiffness $k_i = \frac{E_i A_i}{l_i}$, for $i = 1, 2, \ldots, n + 1$.

Example 6.3 Let us consider a vertically hanging rod [4] subjected to self-weight as given in Figure 6.9 having parameter values $E = 2 \times 10^{11}\,\text{N/m}^2$, $\rho = 7800\,\text{kg/m}^3$, $A = 30 \times 10^{-6}\,\text{m}^2$, and $L = 1\,\text{m}$.

Solution Using the FEM for five element, the system of equations is obtained as

$$\frac{5EA}{L}\begin{bmatrix} 2 & -1 & & & \\ -1 & 2 & -1 & & \\ & -1 & 2 & -1 & \\ & & -1 & 2 & -1 \\ & & & -1 & 1 \end{bmatrix}\begin{Bmatrix} u_2 \\ u_3 \\ u_4 \\ u_5 \\ u_6 \end{Bmatrix} = \frac{\rho A L g}{5}\begin{Bmatrix} 1 \\ 1 \\ 1 \\ 1 \\ 0.5 \end{Bmatrix} \tag{6.36}$$

By solving Eq. (6.36), the nodal displacements are obtained as $u_2 = 6.8796 \times 10^{-8}$, $u_3 = 1.223 \times 10^{-7}$, $u_4 = 1.6052 \times 10^{-7}$, $u_5 = 1.8346 \times 10^{-7}$, and $u_6 = 1.911 \times 10^{-7}$.

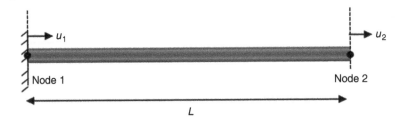

Figure 6.10 Fixed free-bar element.

6.4.2 Dynamic Analysis

In order to incorporate the usage of the FEM in dynamic analysis of structural systems, one-dimensional fixed free bar [2] governed by

$$AE\frac{\partial^2 u}{\partial x^2} = \rho A\frac{\partial^2 u}{\partial t^2} \tag{6.37}$$

(as depicted in Figure 6.10) is taken into consideration.

For free vibration $u = Ue^{i\omega t}$, where U is the nodal displacement and ω is the natural frequency. Using the FEM, Eq. (6.37) gets transformed to eigenvalue problem,

$$\frac{2EA}{L}u_2 = \frac{\omega^2 \rho AL}{6}u_2.$$

Similarly, for n element bar (Figure 6.11), the eigenvalue problem gets reduced to

$$\frac{nEA}{L}\begin{bmatrix} 2 & -1 & & & \\ -1 & 2 & -1 & & \\ & \ddots & \ddots & \ddots & \\ & & -1 & 2 & -1 \\ & & & -1 & 1 \end{bmatrix}\begin{Bmatrix} u_2 \\ u_3 \\ \vdots \\ u_n \\ u_{n+1} \end{Bmatrix} = \frac{\omega^2 \rho AL}{6n}\begin{bmatrix} 4 & 1 & & & \\ 1 & 4 & 1 & & \\ & \ddots & \ddots & \ddots & \\ & & 1 & 4 & 1 \\ & & & 1 & 2 \end{bmatrix}\begin{Bmatrix} u_2 \\ u_3 \\ \vdots \\ u_n \\ u_{n+1} \end{Bmatrix}$$

$$\tag{6.38}$$

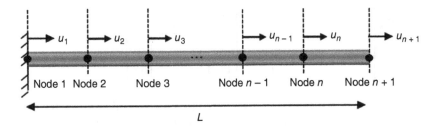

Figure 6.11 Fixed free n element bar.

$$\Rightarrow KU = \omega^2 MU. \tag{6.39}$$

Here, K and M are the stiffness and mass matrices.

Example 6.4 Compute the nodal frequencies for five-element fixed free bar having parameter values $E = 2 \times 10^{11} \, \text{N/m}^2$, $\rho = 7800 \, \text{kg/m}^3$, $A = 30 \times 10^{-6} \, \text{m}^2$, and $L = 1 \, \text{m}$.

Solution By using Eq. (6.38), we obtain the eigenvalue problem after taking five-element assembling,

$$\frac{5EA}{L}\begin{bmatrix} 2 & -1 & & & \\ -1 & 2 & -1 & & \\ & -1 & 2 & -1 & \\ & & -1 & 2 & -1 \\ & & & -1 & 1 \end{bmatrix}\begin{Bmatrix} u_2 \\ u_3 \\ u_4 \\ u_5 \\ u_6 \end{Bmatrix} = \frac{\omega^2 \rho AL}{30}\begin{bmatrix} 4 & 1 & & & \\ 1 & 4 & 1 & & \\ & 1 & 4 & 1 & \\ & & 1 & 4 & 1 \\ & & & 1 & 2 \end{bmatrix}\begin{Bmatrix} u_2 \\ u_3 \\ u_4 \\ u_5 \\ u_6 \end{Bmatrix} \tag{6.40}$$

Solving Eq. (6.40) by substituting the parameter values, we obtain the natural frequencies at nodes u_2, u_3, u_4, u_5, and u_6 as 7 987, 24 752, 43 853, 65 760, and 86 581, respectively.

Exercise

1 Solve boundary value problem $u'' - 9u = 0$, $u(0) = 0$, and $u(1) = 10$ using the Galerkin FEM.

2 Compute the nodal frequencies for 10-element fixed free bar of unit length with material properties Young's modulus $E = 2 \times 10^{11} \, \text{N/m}^2$, density $\rho = 7800 \, \text{kg/m}^3$, and cross-sectional area $A = 30 \times 10^{-6} \, \text{m}^2$.

References

1 Zienkiewicz, O.C., Taylor, R.L., and Zhu, J.Z. (2005). *The Finite Element Method: Its Basis and Fundamentals*, 6e, vol. 132, 1987–1993. Barcelona: Butterworth-Heinemann.

2 Petyt, M. (2010). *Introduction to Finite Element Vibration Analysis*. New York: Cambridge University Press.

3 Rao, S.S. (2017). *The Finite Element Method in Engineering*, 6e. Oxford: Butterworth-Heinemann.

4 Seshu, P. (2003). *Textbook of Finite Element Analysis*. New Delhi: PHI Learning Private Limited.

5 Bhavikatti, S.S. (2005). *Finite Element Analysis*. New Delhi: New Age International.

6 Logan, D.L. (2015). *A First Course in the Finite Element Method*, 6e. Boston: Cengage Learning.

7 Nayak, S. and Chakraverty, S. (2018). *Interval Finite Element Method with MATLAB*. London: Academic Press.

8 Gerald, C.F. and Wheatley, P.O. (2004). *Applied Numerical Analysis*, 7e. New Delhi: Pearson Education.

9 Hoffman, J.D. and Frankel, S. (2001). *Numerical Methods for Engineers and Scientists*, 2e. New York: McGraw-Hill, Inc.

10 Jain, M.K. (2014). *Numerical Solution of Differential Equations*. New Delhi: New Age International.

11 Evans, G.A., Blackledge, J.M., and Yardley, P.D. (eds.) (2000). Finite element method for ordinary differential equations. In: *Numerical Methods for Partial Differential Equations*, 123–164. London: Springer.

12 Kattan, P.I. (2010). *MATLAB Guide to Finite Elements: An Interactive Approach*. New York: Springer Science and Business Media.

7

Finite Volume Method

7.1 Introduction

Finite volume method (FVM) is another widely used numerical technique [1, 2]. The fundamental conservation property of the FVM makes it the preferred method compared to various existing methods viz. finite difference method (FDM), finite element method (FEM), etc. In this approach, similar to the known numerical methods like FDM (Chapter 5) or FEM (Chapter 6), the *volumes* (elements or cells) are evaluated at discrete places over a meshed geometry. Then, the involved volume integrals of respective differential equation containing divergence term are converted to surface integrals by using well-known divergence theorem [3, 4]. Further, the simulated differential equation gets transformed over differential volumes into discrete system of algebraic equations. In the final step, the system of algebraic equations is solved by standard methods to determine the dependent variables.

Aforementioned FVM has various other advantages in handling differential equations occurring in science and engineering problems. For instance, a key feature of the FVM is formulation of physical space (domain of the differential equation) on unregulated polygonal meshes. Another feature is that the FVM is quite easy to implement various boundary conditions in a noninvasive manner, because the involved unknown variables are calculated at the centroids of the volume elements, rather than at their boundary faces [5]. These characteristics of the FVM have made it suitable for the numerical simulations in variety of applications viz. fluid flow, heat and mass transfer, etc. [6, 7]. As such, in this chapter a brief background of the FVM is explained with respect to a simple example problem.

In the next section, we present the discretization techniques of the FVM.

Advanced Numerical and Semi-Analytical Methods for Differential Equations, First Edition.
Snehashish Chakraverty, Nisha Rani Mahato, Perumandla Karunakar, and Tharasi Dilleswar Rao.
© 2019 John Wiley & Sons, Inc. Published 2019 by John Wiley & Sons, Inc.

7.2 Discretization Techniques of FVM

In the FVM, the considered domain on which the conservation laws are applied is subdivided into a set of nonoverlapping cells [4, 5]. The conservation laws are applied on each cell to regulate the variables at some discrete points named as *nodes*. Similar to the FEM [8], there is a considerable freedom in the choice of cells and nodes (i.e. nodes may be at various locations of the cells, such as cell-centers, cell-vertices, or mid-sides). As such, the complete geometrical approach of the FEM can be used in the FVM. There exist different types of cells viz. triangular, quadrilateral, etc. In this regard, Figure 7.1 shows some typical examples of choices of nodes with the associated definition variables [1, 2].

The following section presents the general solving procedure of a convection–diffusion problem using the FVM.

7.3 General Form of Finite Volume Method

The FVM depends on approximate solution of the integral form with respect to conservation equations [2, 9, 10]. In the FVM, the given domain of differential equation is divided into a set of nonoverlapping finite volumes and then the respective integrals of the conservation equations are evaluated by using nodal (function) values at computational nodes. For easy understanding of readers, we present the FVM procedure by solving the convection–diffusion problem. The control volume domain of the convection–diffusion problem is presented in Figure 7.2.

The general form of convection–diffusion problem [9, 11] is represented as

$$\frac{\partial(\rho\phi)}{\partial t} + \nabla \cdot (\rho u\phi) = \nabla \cdot (D\nabla\phi) + S_\phi \tag{7.1}$$

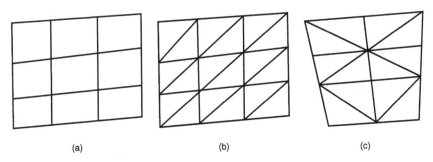

(a)   (b)   (c)

Figure 7.1 Different discretized grids: (a) structured quadrilateral, (b) structured triangular, and (c) unstructured triangular.

Figure 7.2 Sketch of the control volume V and its bounding surface S.

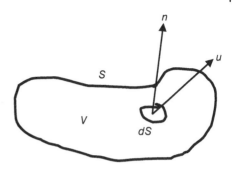

where

$\dfrac{\partial(\rho\phi)}{\partial t}$ = Transient term (ρ-Density, ϕ-Conservative form of all fluid flow),

$\nabla \cdot (\rho u\phi)$ = Convection term (u-velocity),

$\nabla \cdot (D\nabla\phi)$ = Diffusion term (D-Diffusion coefficient) and S_ϕ = Source term.

Integrating Eq. (7.1) over the given domain V, we obtain

$$\int_V \frac{\partial(\rho\phi)}{\partial t}dV + \int_V \nabla \cdot (\rho u\phi)dV = \int_V \nabla \cdot (D\nabla\phi)dV + \int_V S_\phi dV \qquad (7.2)$$

For a fixed domain, $\frac{\partial}{\partial t}$ and $\int_V(\rho\phi)dV$ commute each other [11], so we have

$$\int_V \frac{\partial(\rho\phi)}{\partial t}dV = \frac{\partial}{\partial t}\int_V \rho\phi dV \qquad (7.3)$$

Then, using the Gauss-divergence theorem [12], we change the volume integrals of Eq. (7.2) into surface integrals as

$$\int_V \nabla \cdot (\rho u\phi)dV = \int_S n \cdot (\rho u\phi)dS \qquad (7.4)$$

$$\int_V \nabla \cdot (D\nabla\phi)dV = \int_S n \cdot (D\nabla\phi)dS \qquad (7.5)$$

where n-is the outward unit normal of the surface S (Fig. 7.2).

Now substituting Eqs. (7.3)–(7.5) in Eq. (7.2), the required integral form for the FVM is obtained as

$$\frac{\partial}{\partial t}\int_V \rho\phi dV + \int_S n \cdot (\rho u\phi)dS = \int_S n \cdot (D\nabla\phi)dS + \int_V S_\phi dV \qquad (7.6)$$

7.3.1 Solution Process Algorithm

This section discusses the steps required for solving the convection–diffusion problem using the FVM.

Step (i): Discretize the given domain as a set of *finite volumes* and define the computational nodes to evaluate the required variables.

Step (ii): Apply integral form of conservation law to each finite volume.

Step (iii): Estimate the surface and volume integrals using quadrature formulae at computational nodes.

Step (iv): Assemble all the algebraic equations of the finite volumes to obtain a system of algebraic equations.

Step (v): Solve the obtained system of algebraic equations to find unknown values of the variables at computational nodes.

In Section 7.4, we implement the above steps to solve steady-state convection–diffusion problem.

7.4 One-Dimensional Convection–Diffusion Problem

The steady-state convection–diffusion equation [9, 11] can be derived from transport equation (7.1) by replacing transient term $\left(\dfrac{\partial(\rho\phi)}{\partial t} = 0 \right)$ as

$$\nabla \cdot (\rho u \phi) = \nabla \cdot (D\nabla \phi) + S_\phi. \tag{7.7}$$

Then, the grid generation for solving Eq. (7.7) is illustrated in Section 7.4.1.

7.4.1 Grid Generation

In the initial step of grid generation, divide the considered domain into equal parts of small domain and place nodal points at the midway in between each small domain as given in Figure 7.3. Then, using these nodal points, create control volume near the edge in such a way that the physical boundaries coincide with control volume boundaries [5, 6].

Further, place a general nodal point P for the control volume on the space between A and B in Figure 7.3. The boundaries of control volumes are placed in the middle, between the adjacent nodes. Then, each node is enclosed by a control volume or cell. The nodes to the west and east are identified by W and E, respectively. The west-side face and east-side face of control volume are referred to as w and e, respectively. The distances between WP, wp, pe, and PE are denoted by δx_{WP}, δx_{wp}, δx_{pe}, and δx_{PE}, respectively.

7.4.2 Solution Procedure of Convection–Diffusion Problem

In the absence of source S_ϕ from Eq. (7.7), the steady-state convection–diffusion in one-dimensional flow field [11, 13] is given by

$$\frac{d}{dx}(\rho\phi u) = \frac{d}{dx}\left(D\frac{d\phi}{dx} \right) \tag{7.8}$$

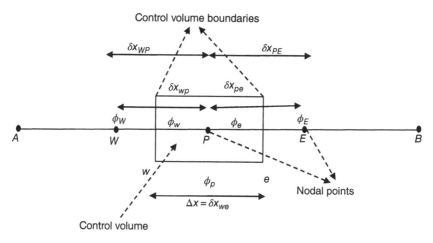

Figure 7.3 Grid generation by dividing the domain into discrete control volumes.

Then, by integrating Eq. (7.8) as $\int_S \frac{d}{dx}(\rho\phi u)ds = \int_S \frac{d}{dx}\left(D\frac{d\phi}{dx}\right)ds$ over control volume [11, 13], we may obtain

$$(\rho uS\phi)_e - (\rho uS\phi)_w = \left(DS\frac{d\phi}{dx}\right)_e - \left(DS\frac{d\phi}{dx}\right)_w \tag{7.9}$$

where the gradient $\left(\frac{d\phi}{dx}\right)$ from east to west is calculated with the help of nodal points (Figure 7.3) as

$$\left.\begin{array}{l} \left(\frac{d\phi}{dx}\right)_e = \frac{\phi_E - \phi_P}{\delta x_{PE}}, \\[2mm] \left(\frac{d\phi}{dx}\right)_w = \frac{\phi_P - \phi_W}{\delta x_{WP}}. \end{array}\right\} \tag{7.10}$$

Now, by substituting Eq. (7.10) in Eq. (7.9), we may get

$$(\rho uS\phi)_e - (\rho uS\phi)_w = (DS)_e\left(\frac{\phi_E - \phi_P}{\delta x_{PE}}\right) - (DS)_w\left(\frac{\phi_P - \phi_W}{\delta x_{WP}}\right). \tag{7.11}$$

Assume $S_w = S_e = S$, $\delta x_{PE} = \delta x_{WP} = \delta x$, $F = \rho u$ (convective mass flux per unit area), and $D_1 = \frac{D}{\delta x}$ (diffusion conductance) in order to obtain the discretized equation [9, 11]. Further, by rearranging Eq. (7.11), we obtain

$$F_e\phi_e - F_w\phi_w = (D_1)_e(\phi_E - \phi_P) - (D_1)_w(\phi_P - \phi_W) \tag{7.12}$$

where $F_e = (\rho u)_e$ and $F_w = (\rho u)_w$.

The central difference approximation with linear interpolation [9, 13, 14] has been used to compute the cell face values ϕ_e and ϕ_w in a uniform grid

$$\phi_e = \frac{(\phi_P + \phi_E)}{2}, \quad \phi_w = \frac{(\phi_W + \phi_P)}{2}. \tag{7.13}$$

On substituting Eq. (7.13) in Eq. (7.12), we obtain

$$F_e\left(\frac{\phi_P + \phi_E}{2}\right) - F_w\left(\frac{\phi_W + \phi_P}{2}\right) = (D_1)_e(\phi_E - \phi_P) - (D_1)_w(\phi_P - \phi_W).$$

$$(7.14)$$

Rearranging Eq. (7.14), we get

$$\left[\left((D_1)_w + \frac{F_w}{2}\right) + \left((D_1)_e - \frac{F_e}{2}\right) + (F_e - F_W)\right]\phi_P$$
$$= \left((D_1)_w + \frac{F_w}{2}\right)\phi_W + \left((D_1)_e - \frac{F_e}{2}\right)\phi_E \qquad (7.15)$$

After simplifying Eq. (7.15), we obtain the discretized equation as

$$c_p \phi_P = c_W \phi_W + c_E \phi_E \qquad (7.16)$$

where

$$c_W = \left((D_1)_w + \frac{F_w}{2}\right) \qquad (7.17)$$

$$c_E = \left((D_1)_e - \frac{F_e}{2}\right) \qquad (7.18)$$

$$c_p = \left((D_1)_w + \frac{F_w}{2}\right) + \left((D_1)_e - \frac{F_e}{2}\right) + (F_e - F_W) \qquad (7.19)$$

and $F_e = F_W = F \Rightarrow F_e - F_W = 0$.

Now, by using Eqs. (7.17) and (7.18), one can rewrite Eq. (7.19) as

$$c_p = c_W + c_E. \qquad (7.20)$$

Finally, the linear system (discretized equations) (7.16) can be handled using any standard method to evaluate the unknown cell face values ϕ_P, ϕ_W, and ϕ_E.

Example 7.1 Calculate the distribution of ϕ in convection–diffusion equation [9, 11, 13] $\frac{d}{dx}(\rho\phi u) = \frac{d}{dx}\left(D\frac{d\phi}{dx}\right)$, as a function of x by using five equally spaced cells with the boundary conditions $\phi(x = 0) = 1$, $\phi(x = 1) = 0$, and $u = 0.1$ m/s, $\rho = 1.0$ kg/m^3, $D = 0.1$ kg/m.

Solution Let the domain is divided into five control volumes with step length $\delta x = 0.2$ m as presented in Figure 7.4 and by using the data $u = 0.1$ m/s, $\rho = 1.0$ kg/m^3, $D = 0.1$ kg/m, we can calculate $F = \rho u = 1.0$, $D_1 = \frac{D}{\delta x} = 0.5$.

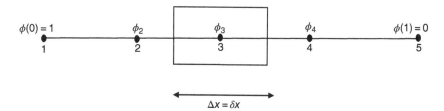

Figure 7.4 Discretization for nodal point 3.

Figure 7.4 also depicts the discretization at nodal point 3. Comparing Figure 7.4 with Figure 7.3, the west face ϕ_W of the cell (nodal point 3) is identified as ϕ_2, the nodal point ϕ_P is identified by ϕ_3, and the east face of the cell ϕ_E is identified by ϕ_4. By applying similar discretization to five nodal points (1, 2, ..., 5), one may get the system of discretized equations. The discretized equations at internal nodal points 2, 3, and 4 can be obtained from Eqs. (7.16)–(7.18) and (7.20), but the nodal points 1 and 5 need special treatment.

The required numerical values at nodes 2, 3, and 4 are obtained using Eqs. (7.16)–(7.18) and (7.20) as

$$c_W = \left(D_1 + \frac{F}{2}\right) = 0.55, \quad c_E = \left(D_1 - \frac{F}{2}\right) = 0.45, \quad \text{and} \quad c_P = c_W + c_E = 1. \tag{7.21}$$

Now, the discretized equations for the fixed intermediate nodes are illustrated below:

1. **Discretized equation for nodal point 2:** At nodal point 2, the west face ϕ_W is identified as ϕ_1, ϕ_P is referred by ϕ_2, and the east face of the cell ϕ_E is represented by ϕ_3, respectively. Now, from Eqs. (7.16) and (7.21), the discretized equation is obtained as

$$c_P\phi_2 = c_W\phi_1 + c_E\phi_3 \Rightarrow \phi_2 - 0.55\phi_1 - 0.45\phi_3 = 0 \tag{7.22}$$

2. **Discretized equation for nodal point 3:** At nodal point 3 (in similar fashion) ϕ_W is identified as ϕ_2, ϕ_P are referred to as ϕ_3, and ϕ_E is represented by ϕ_4. The discretized equation at nodal point 3 is obtained as

$$c_P\phi_3 = c_W\phi_2 + c_E\phi_4 \Rightarrow \phi_3 - 0.55\phi_2 - 0.45\phi_4 = 0 \tag{7.23}$$

3. **Discretized equation for nodal point 4:** Similarly, for nodal point 4, $\phi_W = \phi_3$, $\phi_P = \phi_4$, and $\phi_E = \phi_5$ and also the discretized equation at nodal point 4 may be obtained as

$$c_P\phi_4 = c_W\phi_3 + c_E\phi_5 \Rightarrow \phi_4 - 0.55\phi_3 - 0.45\phi_5 = 0 \tag{7.24}$$

Now, the discretized equation at end nodes is discussed below.

4. **Discretized equation for nodal point 1:** It is a special case because from Figure 7.4 one can observe that the nodal cell 1 does not have the west face. So, by integrating Eq. (7.8) at nodal point 1 and using central difference at the east face of cell 1, we obtain the discretized equation from Eq. (7.15) at nodal point 1 as

$$\frac{F_E}{2}(\phi_P + \phi_E) - F_A \phi_A = (D_1)_E (\phi_E - \phi_P) - (D_1)_A (\phi_P - \phi_A). \qquad (7.25)$$

By using $F_E = F_A = F$, $(D_1)_E = D_1$, $(D_1)_A = 2D_1$, and the values of ϕ at the boundaries given as $\phi_A = 1$ and $\phi_B = 0$, respectively. Then, Eq. (7.25) is transformed to

$$\left(\frac{F}{2} + 3D_1\right)\phi_P + \left(\frac{F}{2} - D_1\right)\phi_E = F + 2D_1. \qquad (7.26)$$

Further, in order to get the discretized equation, the cell faces of nodal point 1 viz. ϕ_P is identified as ϕ_1 and east face ϕ_E is identified by ϕ_2. From Eq. (7.26) the discretized equation for nodal point 1 using the given values of F and D_1 can be obtained as

$$1.55\phi_1 - 0.45\phi_2 = 1.1. \qquad (7.27)$$

5. **Discretized equation for nodal point 5:** Similarly, for control volume 5 given in Figure 7.4 we can observe the nodal cell 5 does not have east face. So, by integrating Eq. (7.8) at nodal point 5 and using central difference scheme at the east face of cell 5, the discretized equation of nodal point 5 from Eq. (7.15) is obtained as

$$F_B \phi_B - \frac{F_W}{2}(\phi_P + \phi_W) = (D_1)_B (\phi_B - \phi_P) - (D_1)_W (\phi_P - \phi_W). \qquad (7.28)$$

Replacing $F_W = F_B = F$, $(D_1)_W = D_1$, $(D_1)_B = 2D_1$, and the values of ϕ at the boundaries given as $\phi_A = 1$ and $\phi_B = 0$ in the Eq. (7.28), we may have

$$\left(3D_1 - \frac{F}{2}\right)\phi_P + \left(-\frac{F}{2} - D_1\right)\phi_W = 0 \qquad (7.29)$$

For the nodal point 5, the cell faces ϕ_P is identified as ϕ_5 and the west face ϕ_W of the nodal point 5 is identified by ϕ_4. From Eq. (7.29) the discretized equation for nodal point 5 using the given values of F and D_1 can be obtained as

$$1.45\phi_5 - 0.55\phi_4 = 0. \qquad (7.30)$$

Finally the required, linear system of algebraic equations obtained from the discrete equations (7.22)–(7.26) for all the nodal points is

$$
\left.\begin{array}{r}
\phi_2 - 0.55\phi_1 - 0.45\phi_3 = 0 \\
\phi_3 - 0.55\phi_2 - 0.45\phi_4 = 0 \\
\phi_4 - 0.55\phi_3 - 0.45\phi_5 = 0 \\
1.55\phi_1 - 0.45\phi_2 = 1.1 \\
1.45\phi_5 - 0.55\phi_4 = 0.
\end{array}\right\} \tag{7.31}
$$

After solving the linear system equation (7.31), we obtain the unknown variables as $\phi_1 = 0.9421$, $\phi_2 = 0.8006$, $\phi_3 = 0.6276$, $\phi_4 = 0.4163$, and $\phi_5 = 0.1579$.

It is worth mentioning here that only the basic idea of FVM has been given for a standard problem. However, interested readers may refer dedicated FVM books [1, 2, 5, 8, 9, 15] for further details.

Exercise

1 Calculate the steady-state temperature in the rod for 1D heat conduction equation. Ends are maintained at constant temperature of 100 and 500 °C with $L = 0.5$ m, respectively and thermal conductivity $K = 1000$ W/mK, cross-sectional area $A = 10 \times 10^{-3}$ m^2, $L = 1$ cm (Hint: Take five control volumes).

2 Calculate the steady-state temperature in a large plate of thickness $L = 2$ cm, constant thermal conductivity $K = 0.5$ W/mK, and uniform heat generation $q = 1000$ kW/m^3. The faces A and B are at temperatures 100 and 200 °C, respectively.

References

1 LeVeque, R.J. (2002). *Finite Volume Methods for Hyperbolic Problems*, vol. 31. Cambridge: Cambridge University Press.
2 Eymard, R., Gallouët, T., and Herbin, R. (2000). Finite volume methods. *Handbook of Numerical Analysis* 7: 713–1018.
3 Godlewski, E. and Raviart, P.A. (2013). *Numerical Approximation of Hyperbolic Systems of Conservation Laws*, vol. 118. New York: Springer Science & Business Media.
4 Raithby, G.D. and Chui, E.H. (1990). A finite-volume method for predicting a radiant heat transfer in enclosures with participating media. *Journal of Heat Transfer* 112 (2): 415–423.

5 Liu, Z.L. (ed.) (2018). Finite volume method. In: *Multiphysics in Porous Materials*, 385–395. Cham: Springer.

6 Demirdžić, I. and Perić, M. (1990). Finite volume method for prediction of fluid flow in arbitrarily shaped domains with moving boundaries. *International Journal for Numerical Methods in Fluids* 10 (7): 771–790.

7 Mingham, C.G. and Causon, D.M. (1998). High-resolution finite-volume method for shallow water flows. *Journal of Hydraulic Engineering* 124 (6): 605–614.

8 Bathe, K.J. and Wilson, E.L. (1976). *Numerical Methods in Finite Element Analysis*. Upper Saddle River: Prentice Hall.

9 Lazarov, R.D., Mishev, I.D., and Vassilevski, P.S. (1996). Finite volume methods for convection-diffusion problems. *SIAM Journal on Numerical Analysis* 33 (1): 31–55.

10 Fuhrmann, J., Ohlberger, M., and Rohde, C. (eds.) (2014). *Finite Volumes for Complex Applications VII-Elliptic, Parabolic and Hyperbolic Problems: FVCA 7, Berlin, June 2014*, Springer Proceedings in Mathematics and Statistics, vol. 78. Cham: Springer.

11 Stynes, M. (1995). Finite volume methods for convection-diffusion problems. *Journal of Computational and Applied Mathematics* 63 (1–3): 83–90.

12 Pfeffer, W.F. (2016). *The Divergence Theorem and Sets of Finite Perimeter*. Boca Raton: Chapman & Hall/CRC.

13 Shukla, A., Singh, A.K., and Singh, P. (2011). A comparative study of finite volume method and finite difference method for convection-diffusion problem. *American Journal of Computational and Applied Mathematics* 1 (2): 67–73.

14 Jasak, H. (1996). Error analysis and estimation for the finite volume method with applications to fluid flows. PhD thesis. Imperial College London.

15 Samarskii, A.A. (2001). *The Theory of Difference Schemes*, Monographs and Textbooks in Pure and Applied Mathematics, vol. 240. New York: Marcel Dekker.

8

Boundary Element Method

8.1 Introduction

In 1977, for the first time the term boundary element method (BEM) was coined in some publications and the first book on the BEM appeared in 1992 [1]. In the last 20 years the BEM has become one of the important numerical computational techniques along with the FDM (Chapter 5) and FEM (Chapter 6). It is used to solve certain classes of differential equations by formulating as integral equations [2, 3]. It is worth mentioning that for certain classes of differential equations, integral equations' reformulation may not always be possible [3, 4]. The advantage in using the BEM is that only the boundaries of the considered domain of the differential equations require subdivision to develop a boundary mesh [5, 6]. Whereas, in other methods like FDM or FEM the whole domain of the differential equation requires discretization. In this chapter, we are presenting a brief introduction of the BEM along with a simple example problem for easy understanding of the method.

8.2 Boundary Representation and Background Theory of BEM

Detailed BEM formulation for handling boundary value problem (BVP) in a field or domain may be found in Refs. [6, 7]. By using known data on the boundary S, one can get the rest of information on the boundary S as well as in the domain D in the BEM. Due to various difficulties [7, 8], BVPs do not always have closed-form solutions or analytic solutions. In general, this may be done by the facility of defining the boundary as a set of panels which have the same characteristic [8]. For example, a closed two-dimensional boundary may be represented by a set of straight lines (panels), as shown Figure 8.1.

Let S be the original boundary and $\Delta \tilde{S}_j$ (for $j = 1, 2, \ldots, n$) be known as the panels that represent an approximation to S. If $\tilde{S} = \sum_{j=1}^{n} \Delta \tilde{S}_j$ is the surface described

Advanced Numerical and Semi-Analytical Methods for Differential Equations, First Edition.
Snehashish Chakraverty, Nisha Rani Mahato, Perumandla Karunakar, and Tharasi Dilleswar Rao.
© 2019 John Wiley & Sons, Inc. Published 2019 by John Wiley & Sons, Inc.

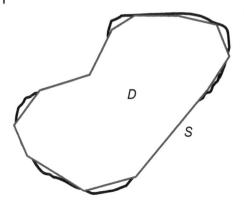

Figure 8.1 Closed two-dimensional boundary represented by a set of straight lines (panels).

D

S

by the complete set of panels, then $\tilde{S}$ is the approximation to ($\tilde{S} \approx S$). The representation of the boundary in this way is the first step to discretize the integral operators that occur in the boundary integral formulations [8, 9].

Next, we address linear differential operator, fundamental solution, and Green's integral theorems which play an important role in the BEM.

8.2.1 Linear Differential Operator

A linear differential operator, denoted by L, is a function [8, 9] such that

In one dimension:

$$L = \frac{d^n}{dx^n} + p_1(x)\frac{d^{n-1}}{dx^{n-1}} + \cdots + p_{n-1}(x)\frac{d}{dx} + p_n(x) \tag{8.1}$$

In two dimensions:

$$L = \frac{\partial^n}{\partial x^n} + p_1(x,y)\frac{\partial^{n-1}}{\partial x^{n-1}} + q_0(x,y)\frac{\partial^n}{\partial y^n} + q_1(x,y)\frac{\partial^{n-1}}{\partial y^{n-1}}$$
$$+ r_1(x,y)\frac{\partial^{n-1}}{\partial x^m \partial y^{n-1-m}} + \cdots + r_{n-2}(x,y)\frac{\partial^2}{\partial x \partial y}$$
$$+ p_{n-1}(x,y)\frac{\partial}{\partial x} + q_{n-1}(x,y)\frac{\partial}{\partial y} + p_n(x,y) \tag{8.2}$$

All linear operators have the property that $L(\alpha u + \beta v) = \alpha L(u) + \beta L(v)$, where α and β are constants and u and v are functions over the domain. The adjoint operator of L represented as L^* is the function defined as

$$\int_D L(u)v dD = \int_D u L^*(v) dD \tag{8.3}$$

and also due to the self-adjoint property we have $L = L^*$.

To use the BEM for solving BVPs, we must transform the problem into an equivalent boundary integral equation (BIE) problem. In this regard, the fundamental solution and Green's integral theorems are very useful tools. Accordingly, we discuss next the fundamental solution, Green's function, and integral theorems.

8.2.2 The Fundamental Solution

Before discussing the fundamental solution, a brief review of the Heaviside function (H) and Dirac delta function δ in R^2 is needed.

8.2.2.1 Heaviside Function
The function $H(x - c)$ is said to be the Heaviside function [10] and is defined as (Chapter 2)

$$H(x - c) = \begin{cases} 0 & \text{if } x < c \\ 1 & \text{if } x \geq c \end{cases}$$

Heaviside functions are also called as step functions.

8.2.2.2 Dirac Delta Function
Let $q(x, y)$ and $p(x_1, y_1)$ represent moving and fixed points [10, 11] in the domain D as presented in Figure 8.2. The set of functions denoted by $\delta_\varepsilon(p, q)$ for $\varepsilon > 0$ is defined as

$$\delta_\varepsilon(p, q) = \delta_\varepsilon(x_1 - x, y_1 - y) = \begin{cases} \dfrac{1}{\varepsilon}, & (x, y) \in D \\ 0, & (x, y) \notin D \end{cases}$$

where ε is the area of the small circle (Figure 8.2).

Figure 8.2 A two-dimensional domain D with boundary S.

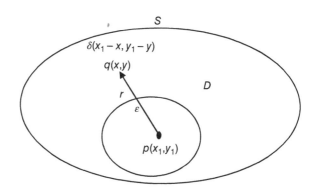

Now, by taking the limit of $\delta_\varepsilon(p, q)$ as $\varepsilon \to 0$ the Dirac delta function δ may be defined as

$$\delta(p, q) = \lim_{\varepsilon \to 0} \delta_\varepsilon(p, q) = \lim_{\varepsilon \to 0} \delta_\varepsilon(x_1 - x, y_1 - y) = \delta(x_1 - x, y_1 - y)$$

$$= \begin{cases} \infty, & (x, y) = (x_1, y_1) \\ 0, & (x, y) \neq (x_1, y_1) \end{cases}.$$

Further, the Dirac delta function can be viewed as the derivative of the Heaviside step function

$$\frac{d}{dx}(H(x)) = \delta(x)$$

and the Dirac delta function has the fundamental property

$$\int_{x-c}^{x+d} \delta(x - s)f(s)ds = f(x).$$

8.2.2.3 Finding the Fundamental Solution

The fundamental solution of the Laplace equation [9, 12] will be described using the Dirac delta function, where the Laplace equation in two dimension is

$$\nabla^2 z = \frac{\partial^2 z}{\partial x^2} + \frac{\partial^2 z}{\partial y^2} = 0 \tag{8.4}$$

The Dirac delta function will be used to derive the fundamental solution of Eq. (8.4), then it should satisfy

$$\nabla^2 w = -\delta(x_1 - x, y_1 - y) \tag{8.5}$$

In polar coordinates the Laplacian operator can easily be written and due to the radial symmetric property [11] we have

$$\nabla^2 = \frac{1}{r}\frac{d}{dr}\left(r\frac{d}{dr}\right) \tag{8.6}$$

where $r = \sqrt{(x_1 - x)^2 + (y_1 - y)^2}$.

Furthermore, the Dirac delta in polar coordinates [12, 13] may be obtained as

$$\delta(x_1 - x, y_1 - y) = \frac{\delta(r)}{2\pi r} \tag{8.7}$$

Hence, from Eqs. (8.5)–(8.7) we have

$$\frac{1}{r}\frac{d}{dr}\left(r\frac{d}{dr}w\right) = -\frac{\delta(r)}{2\pi r} \tag{8.8}$$

By applying multiple integration on both sides of Eq. (8.8) over the domain D, one may obtain the fundamental solution as

$$w(r) = -\frac{1}{2\pi}\log(r) + c_1 \log r + c_2 \tag{8.9}$$

where c_1 and c_2 are integral constants. By considering $c_1 = c_2 = 0$, we may have

$$w(r) = -\frac{1}{2\pi} \log(r) \tag{8.10}$$

8.2.3 Green's Function

Green's function $G(x, s)$ is described as integral kernel that can be used to solve differential equations [12]. Here, the Green's function $G(x, s)$ may be considered as a fundamental solution for the differential equation

$$Lu(x) = f(x) \tag{8.11}$$

So, if $G(x, s)$ is a fundamental solution of Eq. (8.11), then it should satisfy $L[G(x, s)] = -\delta(x - s)$ and the solution $u(x)$ of Eq. (8.11) can be expressed in the following form [12, 13]:

$$u(x) = \int_S -G(x, s)f(s)ds \tag{8.12}$$

From Eqs. (8.11) and (8.12), one can get

$$Lu(x) = L\left[\int_S -G(x, s)f(s)ds\right] \tag{8.13}$$

Since L is a linear operator and $L[G(x, s)] = -\delta(x - s)$, from Eq. (8.13) we have

$$\int_S -L[G(x, s)]f(s)ds = \int_S \delta(x - s)f(s)ds \tag{8.14}$$

Further, by using the known property $\int_{x-c}^{x+d} \delta(x - s)f(s)ds = f(x)$ for any c, $d > 0$, one can rewrite Eq. (8.14) as

$$\int_S -L[G(x, s)]f(s)ds = f(x) \tag{8.15}$$

8.2.3.1 Green's Integral Formula

Green's integral formula is a very useful tool [14] for finding the derivation of integral equations formed in the BEM. Suppose u and v are continuous functions in a domain $D \subset R^2$ with continuous first and second derivatives in D. Then, functions u and v satisfy the Green's formula

$$\int_D (u\nabla^2 v - v\nabla^2 u)dD = \oint_S \left(u\frac{\partial v}{\partial x} - v\frac{\partial u}{\partial x}\right)dS \tag{8.16}$$

where S is the boundary of the domain D and $\frac{\partial u}{\partial x}$ and $\frac{\partial v}{\partial x}$ are the outward normal derivatives of u and v, respectively, on the boundary of domain D.

8.3 Derivation of the Boundary Element Method

The BEM can be derived for differential equations for which we can find first a fundamental solution. Accordingly, here a Laplace equation is handled by using the BEM [15, 16]

$$\nabla^2 z = \frac{\partial^2 z}{\partial x^2} + \frac{\partial^2 z}{\partial y^2} = 0 \tag{8.17}$$

In the BEM the fundamental solution will be used as the weight function in the integral equation. To find a weak fundamental solution, we multiply Eq. (8.17) by a weight function (fundamental solution) w and integrating over the domain D we have

$$\int_D (\nabla^2 z) w \, dD = 0 \tag{8.18}$$

Using the Green–Gauss theorem [17] and Eq. (8.18), one may obtain

$$\int_D (\nabla^2 z) w \, dD = \int_S \frac{\partial z}{\partial n} w \, dS - \int_D \nabla z \nabla w \, dD \tag{8.19}$$

Integrating Eq. (8.19) by parts along with Green's formula, we have

$$\int_D (\nabla^2 z) w \, dD = 0 = \int_S \frac{\partial z}{\partial n} w \, dS - \int_S \frac{\partial w}{\partial n} n \, dS + \int_D z \nabla^2 w \, dD \tag{8.20}$$

It may be noted that in the FEM we chose simple piecewise polynomials as our weighting (test) functions. In the BEM we choose the fundamental solution (Eq. 8.5) so that the last term of Eq. (8.20) becomes

$$\int_D z \nabla^2 w \, dD = -\int_D z \delta(x_1 - x, y_1 - y) \, dD = -z(x_1, y_1) \tag{8.21}$$

Assume that $(x_1, y_1) \in D$ and not on the boundary. From Eqs. (8.18), (8.19), and (8.21), one may obtain

$$\int_S \frac{\partial z}{\partial n} w \, dS - \int_S \frac{\partial w}{\partial n} n \, dS - z(x_1, y_1) = 0 \tag{8.22}$$

8.3.1 BEM Algorithm

This section discusses the solving procedure of the BEM.

Step 1: Find the fundamental solution of the governing equation (Eq. (8.10)).

Step 2: Derive a formula of two field variables which converts domain integrals into boundary integrals. Integrating by parts twice the given equation (Eq. (8.19)).

Step 3: Obtain the BIE for a domain point (Eq. (8.22)).

Step 4: Assume that x approach to the boundary and obtain the BIE for a boundary point.

Step 5: Choose a finite number of boundary points, discretize the BIEs for the boundary points, and put the discretized BIEs in matrix form.

Step 6: In the final step, with respect to the boundary conditions solve the obtained matrix equation.

Next, we present a simple example problem to solve it using the BEM for easy understanding.

Example 8.1   Solve $\frac{d^2z(x)}{dx^2} = 0, \forall x \in (0,1)$ by using the BEM with respect to boundary conditions:

1. Dirichlet boundary conditions $z(0) = 50$, $z(1) = 0$,
2. Neumann boundary conditions $z'(0) = 50$, $z'(1) = 50$,
3. Mixed boundary conditions $z(0) = 50$, $z'(1) = 0$.

Solution   Consider the given problem

$$\frac{d^2z(x)}{dx^2} = 0, \quad \forall x \in (0,1) \tag{8.23}$$

Equation (8.23) has closed-form (or analytical) solutions for three cases of boundary conditions as

Case 1: Closed-form solution of Eq. (8.23) for Dirichlet boundary conditions $z(0) = 50$, $z(1) = 0$.

$$z(x) = 50 - 50x. \tag{8.24}$$

Case 2: For Neumann boundary conditions $z'(0) = 50$, $z'(1) = 50$.

$$z(x) = 50x + c. \tag{8.25}$$

where c is an arbitrary constant.

Case 3: Similarly for Mixed boundary conditions $z(0) = 50$, $z'(1) = 0$.

$$z(x) = 50. \tag{8.26}$$

Solving the Given Problem Eq. (8.23) by Using BEM

Step 1: Let us first find the fundamental solution $Z(x, x_1)$ of the given Eq. (8.23) as below. Let $Z(x, x_1)$ is a fundamental solution, then from Eq. (8.23) it should satisfy

$$\frac{d^2Z(x, x_1)}{dx^2} = -\delta(x - x_1) \tag{8.27}$$

where $\delta(x - x_1)$ is the Dirac delta function.

One can rearrange Eq. (8.27) as

$$\frac{d}{dx}(H(x - x_1)) = \delta(x - x_1) \tag{8.28}$$

By comparing Eqs. (8.27) and (8.28), one can find a solution of Eq. (8.27) as

$$\frac{dZ(x, x_1)}{dx} = -H(x - x_1) + c \tag{8.29}$$

where c is an arbitrary constant and for convenience we set $c = \frac{1}{2}$ and where $H(x - x_1)$ is Heaviside's unit step function [10].

After integrating Eq. (8.29) and choosing the new integration constant as zero, one may obtain the fundamental solution $Z(x, x_1)$ as

$$Z(x, x_1) = \frac{1}{2}x - xH(x - x_1) = -\frac{1}{2}|x - x_1| \tag{8.30}$$

Now, from Eq. (8.30)

$$\frac{\partial Z(x, x_1)}{\partial x} = \frac{1}{2} - H(x - x_1) \tag{8.31}$$

Step 2: Deriving a formula of two field variables which converts domain integrals into boundary integrals. Integrating by parts twice the given Eq. (8.23) by using Eq. (8.20) we get

$$\int_0^1 z\frac{d^2Z}{dx^2}dx = \left(z\frac{\partial Z}{\partial x} - Z\frac{\partial z}{\partial x}\right)_0^1 + \int_0^1 Z\frac{d^2z}{dx^2}dx \tag{8.32}$$

Step 3: Obtaining the BIE for a domain point x_1. By substituting Eqs. (8.23) and (8.31) in Eq. (8.32) yields

$$-z(x) = \left(z(x_1)\frac{\partial Z(x_1, x)}{\partial x} - Z(x_1, x)\frac{\partial z(x_1)}{\partial x}\right)_0^1 \tag{8.33}$$

Next, by introducing the symbols $z' = \frac{\partial z}{\partial x}$ and $W' = \frac{\partial Z}{\partial x}$ in Eq. (8.33), one may find the BIE as

$$-z(x) = z(1)W'(1, x) - Z(1, x)z'(1) - z(0)W'(0, x) + Z(0, x)z'(0). \tag{8.34}$$

Step 4: Let x approach to the boundary and obtain the BIE for a boundary point. Accordingly, as x approaches to the left boundary point zero $(x \to 0)$, Eqs. (8.30) and (8.31) transformed to

$$\begin{cases} \lim_{x \to 0} Z(0, x) = 0, & \lim_{x \to 0} Z(1, x) = \frac{1}{2} \\ \lim_{x \to 0} W'(0, x) = \frac{1}{2}, & \lim_{x \to 0} W'(1, x) = -\frac{1}{2} \end{cases} \tag{8.35}$$

Similarly, as x approaches to the right boundary point one ($x \to 1$), Eqs. (8.30) and (8.31) transformed to

$$\begin{cases} \lim_{x \to 1} Z(0, x) = -\frac{1}{2}, & \lim_{x \to 1} Z(1, x) = 0, \\ \lim_{x \to 1} W'(0, x) = \frac{1}{2}, & \lim_{x \to 1} W'(1, x) = -\frac{1}{2} \end{cases} \tag{8.36}$$

From Eqs. (8.34) and (8.35), one can obtain the BIE for the boundary point $x = 0$ as

$$-z(0) = z(1) \left(\frac{-1}{2} \right) - \left(\frac{1}{2} \right) z'(1) - z(0) \left(\frac{1}{2} \right) \tag{8.37}$$

Again for the boundary point $x = 1$, the BIE may be obtained from Eqs. (8.34) and (8.36) as

$$-z(1) = z(1) \left(\frac{-1}{2} \right) - z(0) \left(\frac{1}{2} \right) - \left(\frac{1}{2} \right) z'(0) \tag{8.38}$$

Step 5: The discretized BIEs in matrix form
By arranging Eqs. (8.37) and (8.38) we have the matrix form as

$$\begin{bmatrix} 0.5 & -0.5 \\ -0.5 & 0.5 \end{bmatrix} \begin{bmatrix} z(0) \\ z(1) \end{bmatrix} = \begin{bmatrix} 0 & -0.5 \\ 0.5 & 0 \end{bmatrix} \begin{bmatrix} z'(0) \\ z'(1) \end{bmatrix} \tag{8.39}$$

Step 6: Solving the above matrix equation for given three cases of boundary conditions
Case 1: By using Dirichlet boundary conditions in Eq. (8.39), we get

$$\begin{bmatrix} 0.5 & -0.5 \\ -0.5 & 0.5 \end{bmatrix} \begin{bmatrix} 50 \\ 0 \end{bmatrix} = \begin{bmatrix} 0 & -0.5 \\ 0.5 & 0 \end{bmatrix} \begin{bmatrix} z'(0) \\ z'(1) \end{bmatrix} \tag{8.40}$$

From Eq. (8.40), one can obtain the solution as

$$\begin{bmatrix} z'(0) \\ z'(1) \end{bmatrix} = \begin{bmatrix} -50 \\ -50 \end{bmatrix} \tag{8.41}$$

After integrating Eq. (8.41), one may obtain

$$\begin{cases} z(0) = -50x + c_1 \\ z(1) = -50x + c_2 \end{cases}$$

By using the given Dirichlet boundary conditions $z(0) = 50$, $z(1) = 0$, one may have the solution as $z(x) = 50 - 50x$ which exactly matches with closed-form Solution (8.24).

Case 2: By using Neumann boundary conditions in Eq. (8.39), the matrix form becomes

$$\begin{bmatrix} 0.5 & -0.5 \\ -0.5 & 0.5 \end{bmatrix} \begin{bmatrix} z(0) \\ z(1) \end{bmatrix} = \begin{bmatrix} 0 & -0.5 \\ 0.5 & 0 \end{bmatrix} \begin{bmatrix} 50 \\ 50 \end{bmatrix} \tag{8.42}$$

Solution of Eq. (8.42) may be found as

$$\begin{bmatrix} z(0) \\ z(1) \end{bmatrix} = \begin{bmatrix} c \\ 50 + c \end{bmatrix} \tag{8.43}$$

By using the given $z'(0) = 50$, $z'(1) = 50$ one may obtain the solution as $z(x) = 50x + c$ which again agrees with the closed-form Solution (8.25).

Case 3: Similarly, for Mixed boundary conditions, from Eq. (8.39) one may have

$$\begin{bmatrix} 0.5 & -0.5 \\ -0.5 & 0.5 \end{bmatrix} \begin{bmatrix} 50 \\ z(1) \end{bmatrix} = \begin{bmatrix} 0 & -0.5 \\ 0.5 & 0 \end{bmatrix} \begin{bmatrix} z'(0) \\ 0 \end{bmatrix} \tag{8.44}$$

The corresponding solution may then be written as

$$\begin{bmatrix} z'(0) \\ z(1) \end{bmatrix} = \begin{bmatrix} 0 \\ 50 \end{bmatrix} \tag{8.45}$$

Again by using Mixed boundary conditions given as $z(0) = 50$, $z'(1) = 0$, one can get the solution as $z(x) = 50$. This is also exactly the same as the closed-form Solution (8.26).

Exercise

1 Solve the following equations by using the BEM:
   (i) Poisson equation
   (ii) Diffusion equation
   (iii) Wave equation
   (iv) Heat equation

References

1 Brebbia, C.A. and Abascal, J.D. (1992). *Dominguez Boundary Elements: An Introductory Course*. Boston: Computational Mechanics Publications.

2 Gaul, L., Kogl, M., and Wagner, M. (2003). *Boundary Element Methods for Engineers and Scientists: An Introductory Course with Advanced Topics*. Berlin: Springer.

3 Hunter, P. and Pullan, A. (1997). *FEM/BEM Notes*. New Zealand: University of Auckland.

4 Trevelyan, J. (1994). *Boundary Elements for Engineers: Theory and Applications*. Boston: Computational Mechanics Publications.

5 Hall, W.S. (ed.) (1994). Boundary element method. In: *The Boundary Element Method*, 61–83. Dordrecht: Springer.

6 Becker, A.A. (1992). *The Boundary Element Method in Engineering: a Complete Course*. Singapore: McGraw-Hill.

7 Partridge, P.W. and Brebbia, C.A. (eds.) (2012). *Dual Reciprocity Boundary Element Method*. Southampton, Boston: Springer Science & Business Media.

8 Banerjee, P.K. and Butterfield, R. (1981). *Boundary Element Methods in Engineering Science*, vol. 17, 37. London: McGraw-Hill.

9 París, F. and Cañas, J. (1997). *Boundary Element Method: Fundamentals and Applications*, vol. 1. Oxford: Oxford University Press.

10 Bracewell, R. (2000). *The Fourier Transform and Its Applications*, 61–65. New York: McGraw-Hill.

11 Khelashvili, A. and Nadareishvili, T. (2017). Dirac's reduced radial equations and the problem of additional solutions. *International Journal of Modern Physics E* 26 (7): 1750043.

12 Berger, J.R. and Tewary, V.K. (2001). Greens functions for boundary element analysis of anisotropic bimaterials. *Engineering Analysis with Boundary Elements* 25 (4–5): 279–288.

13 Cheng, A.D., Abousleiman, Y., and Badmus, T. (1992). A Laplace transform BEM for axisymmetric diffusion utilizing pre-tabulated Green's function. *Engineering Analysis with Boundary Elements* 9 (1): 39–46.

14 Bochner, S. (1943). *Analytic and Meromorphic Continuation by Means of Green's Formula*, Annals of Mathematics, 652–673. Providence, RI: American Mathematical Society.

15 Lesnic, D., Elliott, L., and Ingham, D.B. (1997). An iterative boundary element method for solving numerically the Cauchy problem for the Laplace equation. *Engineering Analysis with Boundary Elements* 20 (2): 123–133.

16 Delvare, F. and Cimetiere, A. (2008). A first order method for the Cauchy problem for the Laplace equation using BEM. *Computational Mechanics* 41 (6): 789–796.

17 Solecki, R. (1983). Bending vibration of a simply supported rectangular plate with a crack parallel to one edge. *Engineering Fracture Mechanics* 18 (6): 1111–1118.

9

Akbari–Ganji's Method

9.1 Introduction

There exist a variety of analytical and numerical methods for solving linear differential equations as discussed in Chapters 1–8, but methods for solving nonlinear differential equations [1, 2] are quite challenging. In this regard, this chapter focuses on an innovative algebraic approach proposed by Akbari et al. [3] for solving nonlinear differential equations. However, in Ref. [4], it is also termed as a semi-analytic method. The algebraic approach was initially referred to as Algebraic Method and then renamed as Akbari–Ganji's method (AGM). Akbari et al. [5] gave a detailed discussion on solving nonlinear, non-vibrational, vibrational, and integro-differential equations using the AGM along with Maple codes.

There exist various other semi-analytical approaches for solving nonlinear differential equations viz. Lyapunov's small parameter method, Adomian decomposition method (ADM), etc. Later in Chapter 11, the ADM approach is discussed for solving nonlinear (ordinary or partial) differential equations. There also exist other new techniques viz. homotopy perturbation method (HPM), variational iteration method (VIM), homotopy analysis method (HAM), etc. which are further considered in Chapters 12–14. In this chapter we will mainly focus on solving nonlinear ordinary differential equations only using the AGM. The chapter is organized such that the next section consists of preliminaries and AGM approach for solving nonlinear ordinary differential equations. Then, numerical examples of unforced and forced nonlinear differential equations have been incorporated into Section 9.3.

Modeling of nonlinear differential equations analytically is rather more difficult compared to solving linear differential equations. In this regard, the AGM may be considered as a powerful algebraic (semi-analytic) approach for solving such problems. In the AGM, initially a solution function consisting of unknown constant coefficients is assumed satisfying the differential equation and the initial conditions (IC). Then, the unknown coefficients are computed using algebraic equations obtained with respect to IC and their derivatives.

Advanced Numerical and Semi-Analytical Methods for Differential Equations, First Edition.
Snehashish Chakraverty, Nisha Rani Mahato, Perumandla Karunakar, and Tharasi Dilleswar Rao.
© 2019 John Wiley & Sons, Inc. Published 2019 by John Wiley & Sons, Inc.

9.2 Nonlinear Ordinary Differential Equations

The basic notion of nonlinear differential equation and its solution procedure is illustrated in this section.

9.2.1 Preliminaries

Let us consider a general form of nonlinear ordinary differential equation

$$A(u) = f(t) \tag{9.1}$$

having $A(u) = L(u) + N(u)$, where $L(u)$ is the linear part, $N(u)$ is the nonlinear part, and $f(t)$ is the applied external force. In the absence of external force, that is when $f(t) = 0$, Eq. (9.1) gets converted to unforced nonlinear differential equation

$$A(u) = 0 \tag{9.2}$$

In Eqs. (9.1) and (9.2), if the highest order of the derivative is of second order, then such problems are referred to as second-order nonlinear forced and unforced differential equations, respectively. Application problems such as some vibrational systems are governed by second-order nonlinear differential equations. In this context, the detailed illustration of the AGM approach is presented in Section 9.2.2 for solving unforced and forced ordinary nonlinear differential equations.

9.2.2 AGM Approach

In particular, consider a second-order nonlinear ordinary differential equation as

$$F(u''(t), u'(t), u(t), f(\omega_0 t)) = 0 \tag{9.3}$$

where $F(u''(t), u'(t), u(t), f(\omega_0 t)) = A(u) - f(t)$ having IC

$$IC : u(0) = A \quad \text{and} \quad u'(0) = B. \tag{9.4}$$

The basic steps involved in the AGM approach for solving Eq. (9.3) are detailed below:

Step (i): Assume a solution in terms of unknown coefficients. For instance, in case of governing differential equations of unforced vibration systems, the solution is assumed in terms of trigonometric function

$$u(t) = e^{-bt}\{a\cos(\omega t + \varphi)\} \tag{9.5}$$

and in the forced case, the initial solution is assumed as

$$u(t) = e^{-bt}\{a\cos(\omega t + \varphi)\} + d\cos(\omega_0 t + \phi). \tag{9.6}$$

Step (ii): Substitute the assumed solution (9.5) in the *IC* given in Eq. (9.4) to obtain

$$a\cos(\varphi) = A,$$

and

$$-ab\cos(\varphi) - a\omega\sin(\varphi) = B$$

for unforced differential equation. In case of forced differential equation, substitute the assumed solution (9.6) in the *IC* to obtain

$$a\cos(\varphi) + d\cos(\phi) = A$$

and

$$-ab\cos(\varphi) - a\omega\sin(\varphi) - d\omega_0\sin(\phi) = B.$$

Further, it may be noted that this step helps in obtaining two algebraic equations in terms of the unknown coefficients. Accordingly, the remaining algebraic equations are obtained using the next step.

Step (iii): Substitute the assumed solution (9.5) or (9.6) in the governing differential equation (9.3) resulting in

$$H(t) = F(u''(t), u'(t), u(t), f(\omega_0 t)). \tag{9.7}$$

Using *IC*, compute the differential equation (9.7) and its derivatives as

$$\left.\begin{aligned}
H(t) &= F(u''(IC), u'(IC), u(IC), f(\omega_0 IC)) \\
H'(t) &= F(u''(IC), u'(IC), u(IC), f(\omega_0 IC)) \\
H''(t) &= F''(u''(IC), u'(IC), u(IC), f(\omega_0 IC))
\end{aligned}\right\} \tag{9.8}$$

and so on.

It is worth mentioning that the number of equations obtained using steps (ii) and (iii) are equivalent to numbers of unknowns.

9.3 Numerical Examples

This section illustrates the efficiency of the procedure discussed in Section 9.2.2 using unforced and forced nonlinear differential equations with respect to Helmholtz and Duffing equations in Sections 9.3.1 and 9.3.2, respectively.

9.3.1 Unforced Nonlinear Differential Equations

Let us consider an unforced nonlinear ordinary differential equation as

$$F(u''(t), u'(t), u(t)) = 0 \tag{9.9}$$

having *IC*: $u(0) = A$ and $u'(0) = B$. Accordingly, using Eq. (9.5) the solution in terms of four unknown coefficients a, b, ω, and φ is assumed as

$u(t) = e^{-bt}\{a\cos(\omega t + \varphi)\}$. Now, substituting $u(t)$ in Eq. (9.9) gives $F(u''(t), u'(t), u(t)) = 0$, where

$$u'(t) = ae^{-bt}\{b\cos(\omega t + \varphi) - \omega\sin(\omega t + \varphi)\}, \tag{9.10}$$

$$u''(t) = ae^{-bt}\{(b^2 - \omega^2)\cos(\omega t + \varphi) + 2b\omega\sin(\omega t + \varphi)\}. \tag{9.11}$$

Using IC, Eqs. (9.10) and (9.11) get converted to two algebraic equations given by

$$a\cos(\varphi) = A \tag{9.12}$$

and

$$-ab\cos(\varphi) - a\omega\sin(\varphi) = B. \tag{9.13}$$

Further, applying Eq. (9.8), the two more algebraic equations for computing the unknowns a, b, ω, and φ are obtained as

$$F(u''(IC), u'(IC), u(IC)) = 0 \tag{9.14}$$

and

$$F'(u''(IC), u'(IC), u(IC)) = 0. \tag{9.15}$$

Example 9.1 Let us consider Helmholtz equation (Momani et al. [7]) given by

$$y'' + 2y + y^2 = 0, \tag{9.16}$$

having IC

$$y(0) = 0.1 \quad \text{and} \quad y'(0) = 0. \tag{9.17}$$

Solution Assume the solution in terms of unknown coefficients as $y(t) = e^{-bt}\{a\cos(\omega t + \varphi)\}$. Then, using Eqs. (9.12)–(9.15), the algebraic equations are obtained as

$$a\cos(\varphi) = 0.1, \tag{9.18}$$

$$-ab\cos(\varphi) - a\omega\sin(\varphi) = 0, \tag{9.19}$$

$$(ab^2 - a\omega^2 + 2a)\cos(\varphi) + 2ab\omega\sin(\varphi) - a^2\cos^2(\varphi) = 0, \tag{9.20}$$

and

$$(-ab^3 + 3ab\omega^2 - 2ab)\cos(\varphi) + (-3ab^2\omega + a\omega^3 - 2a\omega)\sin(\varphi)$$
$$-2a^2b\cos^2(\varphi) - 2a^2\omega\cos(\varphi)\sin(\varphi) = 0. \tag{9.21}$$

Solving Eqs. (9.18)–(9.21) results in solution set $a = 0.1$, $b = 0$, $\omega = 1.4491$, and $\varphi = 0$. As such, the algebraic solution obtained using the AGM approach is

$$y(t) = 0.1\cos(1.4491t). \tag{9.22}$$

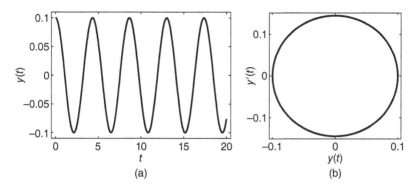

Figure 9.1 Plots of (a) solution $y(t)$ and (b) phase plane of Example 9.1 obtained using the AGM.

Corresponding solution and phase plane plots of Eq. (9.22) have been depicted in Figure 9.1.

In Section 9.3.2, we address forced differential (Duffing) equation using the AGM.

9.3.2 Forced Nonlinear Differential Equation

Let us now consider a forced nonlinear ordinary differential equation as given in Eq. (9.3)

$$F(u''(t), u'(t), u(t), f(\omega_0 t)) = 0$$

having IC

$$IC : u(0) = A \text{ and } u'(0) = B.$$

In case of forced differential equations, the solution in terms of seven unknown coefficients a, b, d, ω, ω_0, φ, and ϕ as given in Eq. (9.6) is assumed as

$$u(t) = e^{-bt}\{a\cos(\omega t + \varphi)\} + d\cos(\omega_0 t + \phi). \tag{9.23}$$

As such, the differential equation may be obtained as $F(u''(t), u'(t), u(t), f(\omega_0 t)) = 0$, where

$$u'(t) = ae^{-bt}\{-b\cos(\omega t + \varphi) - \omega\sin(\omega t + \varphi)\} - d\omega_0\cos(\omega_0 t + \phi) \tag{9.24}$$

and

$$u''(t) = ae^{-bt}\{(b^2 - \omega^2)\cos(\omega t + \varphi) + 2b\omega\sin(\omega t + \varphi)\} - d\omega_0^2\cos(\omega_0 t + \phi). \tag{9.25}$$

Then the two algebraic equations, using *IC* in Eqs. (9.23) and (9.24), are obtained as

$$ab\cos(\varphi) + d\cos(\phi) = A, \qquad (9.26)$$

$$-ab\cos(\varphi) - a\omega\sin(\varphi) - d\omega_0\sin(\phi) = B. \qquad (9.27)$$

Using Eq. (9.8), the remaining five algebraic equations for computing the unknowns a, b, d, ω, ω_0, φ, and ϕ are obtained as

$$\left.\begin{aligned}
&F(u''(IC), u'(IC), u(IC), f(\omega_0 IC)) = 0,\\
&F'(u''(IC), u'(IC), u(IC), f(\omega_0 IC)) = 0,\\
&F''(u''(IC), u'(IC), u(IC), f(\omega_0 IC)) = 0,\\
&F'''(u''(IC), u'(IC), u(IC), f(\omega_0 IC)) = 0,\\
&F^{(iv)}(u''(IC), u'(IC), u(IC), f(\omega_0 IC)) = 0.
\end{aligned}\right\} \qquad (9.28)$$

Above seven algebraic equations (9.26)–(9.28) help in computation of unknown constant coefficients a, b, d, ω, ω_0, φ, and ϕ.

Example 9.2 Let us consider the Duffing equation (Yusufoğlu [8]) given by

$$u''(t) + 3u(t) - 2u^3(t) = \cos(t)\sin(2t) \qquad (9.29)$$

having *IC*

$$u(0) = 0, \quad u'(0) = 1. \qquad (9.30)$$

Solution Assume the solution in terms of unknown coefficients for Eq. (9.29) as $u(t) = e^{-bt}\{a\cos(\omega t + \varphi)\} + d\cos(\omega_0 t + \phi)$. Using Eqs. (9.26)–(9.28), the algebraic equations are obtained as

$$a\cos(\varphi) + d\cos(\phi) = 0, \qquad (9.31)$$

$$-ab\cos(\varphi) - a\omega\sin(\varphi) - d\omega_0\sin(\phi) = 1, \qquad (9.32)$$

$$\begin{aligned}
&(ab^2 - a\omega^2 + 3a)\cos(\varphi) + 2ab\omega\sin(\varphi)\\
&+ (3 - \omega_0^2)d\cos(\phi) - 2(a\cos(\varphi) + d\cos(\phi))^3 = 0,
\end{aligned} \qquad (9.33)$$

$$\begin{aligned}
&(ab^2 - a\omega^2 + 3a)b\cos\varphi + (-3b^2\omega + \omega^3 - 3)\sin\varphi + (\omega_0^2 - 3)d\omega_0\\
&\sin\phi + 6(a\cos\varphi + d\cos\phi)^2(-ab\cos\varphi - a\omega\sin\varphi - d\omega_0\sin\phi) = 2.
\end{aligned} \qquad (9.34)$$

$$\begin{aligned}
&(ab^4 - 6ab^2\omega^2 + a\omega^4 + 3ab^2 - 3a\omega^2)\cos(\varphi) + (4ab^3\omega - 4ab\omega^3 + 6ab\omega)\sin(\varphi)\\
&+ (d\omega_0^4 - 3d\omega_0^2)\cos(\phi) - 12(a\cos(\varphi) + d\cos(\phi))\\
&\times (-ab\cos(\varphi) - a\omega\sin(\varphi) - d\omega_0\sin(\phi))^2 - 6(a\cos(\varphi) + d\cos(\phi))^2\\
&\times (ab^2\cos(\varphi) + 2ab\omega\sin(\varphi) - a\omega^2\cos(\varphi) - d\omega_0^2\cos(\phi)) = 0. \qquad (9.35)
\end{aligned}$$

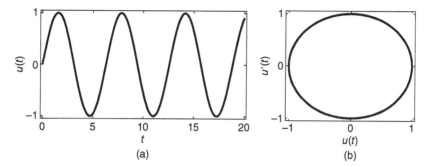

Figure 9.2 Plots of (a) solution $u(t)$ and (b) phase plane of Example 9.2 obtained using the AGM.

It results in a 7×7 system of algebraic equations having seven unknowns resulting in a solution set $a = -1$, $b = 0$, $d = 1$, $\omega = -1$, $\omega_0 = 0$, $\varphi = -\frac{\pi}{2}$, and $\phi = \pi$. As such, the solution obtained using the AGM approach is $u(t) = \sin(t)$. The solution and phase plane plots are depicted in Figure 9.2.

It is worth mentioning that the present chapter gives an overview of implementation of the AGM for solving nonlinear second-order ordinary differential equations. In case of the AGM procedure for solving higher-order differential equations, readers are encouraged to refer the books [4, 5]. Further, in order to solve partial differential equations, Laplace transform method may be combined with the AGM as given in Refs. [5, 6].

Exercise

1 Solve the unforced Duffing equation $u''(t) + 3u(t) - 2u^3(t) = 0$ subject to *IC* $u(0) = 0$ and $u'(0) = 0.5$ using the AGM.

2 Compute the solution of forced Duffing oscillator $u''(t) - 2u^3(t) = t$ subject to *IC* $u(0) = 0$ and $u'(0) = 1$ using the AGM.

References

1 Fucik, S. and Kufner, A. (2014). *Nonlinear Differential Equations*, vol. 2. Prague: Elsevier.

2 Hermann, M. and Seravi, M. (2016). *Nonlinear Ordinary Differential Equations*. New Delhi: Springer.

3 Akbari, M.R., Ganji, D.D., Nimafar, M., and Ahmadi, A.R. (2014). Significant progress in solution of nonlinear equations at displacement of structure

and heat transfer extended surface by new AGM approach. *Frontiers of Mechanical Engineering* 9 (4): 390–401.

4 Sheikholeslami, M. and Ganji, D.D. (2018). *Applications of Semi-Analytical Methods for Nanofluid Flow and Heat Transfer*. Oxford: Elsevier.

5 Akbari, M. (2015). *Nonlinear Dynamic in Engineering by Akbari-Ganji's Method*. Xlibris Corporation.

6 Ganji, D.D. and Talarposhti, R.A. (2017). *Numerical and Analytical Solutions for Solving Nonlinear Equations in Heat Transfer*. Hershey, PA: Technology and Engineering, IGI Global.

7 Momani, S., Erjaee, G.H., and Alnasr, M.H. (2009). The modified homotopy perturbation method for solving strongly nonlinear oscillators. *Computers and Mathematics with Applications* 58 (11–12): 2209–2220.

8 Yusufoğlu, E. (2006). Numerical solution of Duffing equation by the Laplace decomposition algorithm. *Applied Mathematics and Computation* 177 (2): 572–580.

10

Exp-Function Method

10.1 Introduction

In Chapter 9, we have studied Akbari–Ganji's Method (AGM), which provides the algebraic solution of linear and nonlinear differential equations. In the present chapter we study another straightforward and simple method called the exp-function method by which one can find the solution (analytical/semi-analytical) of differential equations. This method was first proposed by He and Wu [1] and was successfully applied to obtain the solitary and periodic solutions of nonlinear partial differential equations. Further, this method was used by many researchers for handling various other equations like stochastic equations [2], system of partial differential equations [3], nonlinear evaluation equation of high dimension [4], difference-differential equation [5], and nonlinear dispersive long-wave equation [6].

10.2 Basics of Exp-Function Method

This method will be illustrated for partial differential equations as ordinary differential equations are straightforward while solving partial differential equations. Let us consider a nonlinear partial differential equation

$$F(u, u_x, u_t, u_{xx}, u_{xt}, u_{tt}, \ldots) = 0 \tag{10.1}$$

to understand briefly the exp-function method [1, 7–10].

Here, subscripts indicate partial differentiation with respect to indicated variables in the subscript.

Next, using traveling wave transformation $u = u(\eta)$, $\eta = kx + \omega t$ (where k and ω are constants) in Eq. (10.1), the partial differential equation may be transformed to a nonlinear ordinary differential equation

$$F(u, u', u'', \ldots) = 0 \tag{10.2}$$

where the prime indicates derivative with respect to η.

Advanced Numerical and Semi-Analytical Methods for Differential Equations, First Edition.
Snehashish Chakraverty, Nisha Rani Mahato, Perumandla Karunakar, and Tharasi Dilleswar Rao.
© 2019 John Wiley & Sons, Inc. Published 2019 by John Wiley & Sons, Inc.

The exp-function method is based on the assumption that the solution of ordinary differential Eq. (10.2) can be expressed as

$$u(\eta) = \frac{\displaystyle\sum_{n=-c}^{d} a_n \exp(n\eta)}{\displaystyle\sum_{n=-p}^{q} b_n \exp(n\eta)} \qquad (10.3)$$

where c, d, p, and q are positive integers yet to be determined, and a_n and b_n are unknown constants.

The values of c and p are determined by balancing exp-functions of the linear term of lowest order in Eq. (10.2) with the lowest-order nonlinear term. Similarly, balancing the exp-functions of highest-order linear term in Eq. (10.2) with the highest-order nonlinear term, values of d and q can be obtained.

Then, we may obtain an equation in terms of $\exp(n\eta)$ by substituting Eq. (10.3) along with the above-determined values of c, d, p, and q into Eq. (10.2). Setting all the coefficients of the various powers of $\exp(n\eta)$ to zero lead to a set of algebraic equations for a_n, b_n, k, and ω.

By solving these algebraic equations, the values of a_n, b_n, k, and ω may be found. Further, by incorporating these values into Eq. (10.3), the solution of the nonlinear PDE (10.1) can be obtained.

10.3 Numerical Examples

We solve now two example problems to make the readers understand the exp-function method.

Example 10.1   Let us solve nonlinear parabolic PDE from Ref. [11] as below

$$u_t = u_{xx} + u + \lambda u^3. \qquad (10.4)$$

Solution   Using the transformation $u = u(\eta)$, $\eta = kx + \omega t$ in Eq. (10.4), it reduces to

$$k^2 u'' + u + \lambda u^3 - \omega u' = 0 \qquad (10.5)$$

where the prime denotes the differentiation with respect to η.

We assume the solution of the ordinary differential equation (10.5) as Eq. (10.3). So, we may assume the solution as

$$u(\eta) = \frac{a_{-c} \exp(-c\eta) + \cdots + a_d \exp(d\eta)}{b_{-p} \exp(-p\eta) + \cdots + b_q \exp(q\eta)} \qquad (10.6)$$

The highest-order linear and nonlinear terms in Eq. (10.5) are u'' and u^3.

Differentiating and simplifying Eq. (10.6) we get

$$u'' = \frac{C_{AN} \exp[(-3p - c)\eta] + \cdots + C_{LN} \exp[(d + 3q)\eta]}{C_{AD} \exp[(-4p)\eta] + \cdots + C_{LD} \exp[(4q)\eta]} \tag{10.7}$$

and

$$u^3 = \frac{D_{AN} \exp(-3c\eta) + \cdots + D_{LN} \exp(3d\eta)}{D_{AD} \exp(-3p\eta) + \cdots + C_{LD} \exp(3q\eta)} \tag{10.8}$$

It may be noted that after differentiation and simplification we will get various terms containing different combinations of a_{-c}, a_d, b_{-p}, ..., etc. which for simplicity we assumed as C_{AN}, ..., C_{LN}, C_{AD}, ..., C_{LD}, D_{AN}, ..., D_{LN}, and D_{AD}, ..., D_{LD}.

The lowest-order exp-function term of Eq. (10.8) is

$$\frac{D_{AN} \exp[(-3c)\eta] + \cdots}{D_{AD} \exp[(-3p)\eta] + \cdots}.$$

Multiplying and dividing by $\exp(-p\eta)$, we get

$$\frac{D_{AN} \exp[(-p - 3c)\eta] + \cdots}{D_{AD} \exp[(-4p)\eta] + \cdots}. \tag{10.9}$$

Similarly, the highest-order exp-function term of Eq. (10.8) is

$$\frac{\cdots + D_{LN} \exp[(3d)\eta]}{\cdots + D_{LD} \exp[(3q)]}.$$

Multiplying and dividing by $\exp(q\eta)$, we get

$$\frac{\cdots + D_{LN} \exp[(3d + q)\eta]}{\cdots + D_{LN} \exp[(4q)]}. \tag{10.10}$$

Balancing the lowest-order exp-functions in Eqs. (10.7) and (10.9), we get

$$-3p - c = -p - 3c$$
$$\Rightarrow p = c.$$

Balancing the highest-order of exp-functions in Eqs. (10.7) and (10.10), we have

$$d + 3q = 3d + q$$
$$\Rightarrow d = q.$$

Here, for simplicity, we consider $p = c = 1$ and $d = q = 1$ [1, 7–10]. Then, the trial solution (10.3) takes the form as

$$u(\eta) = \frac{a_{-1} \exp(-\eta) + a_0 + a_1 \exp(\eta)}{b_{-1} \exp(-\eta) + b_0 + b_1 \exp(\eta)} \tag{10.11}$$

For the case $b_1 \neq 0$, Eq. (10.11) can be simplified further as below

$$u(\eta) = \frac{a \exp(\eta) + b + c \exp(-\eta)}{\exp(\eta) + d + f \exp(-\eta)}. \tag{10.12}$$

Here, $a = \frac{a_1}{b_{-1}}, b = \frac{a_0}{b_{-1}}, c = \frac{a_{-1}}{b_{-1}}, d = \frac{b_0}{b_{-1}}$, and $f = \frac{b_1}{b_{-1}}$ are coefficients to be determined next.

Plugging Eq. (10.12) in ODE (10.5), we have

$$\frac{1}{A} \Big[E_{-3} \exp(-3\eta) + E_{-2} \exp(-2\eta) + E_{-1} \exp(-\eta) + E_0 + E_1 \exp(\eta)$$
$$+ E_2 \exp(2\eta) + E_3 \exp(3\eta) \Big] = 0 \tag{10.13}$$

where $A = [\exp(\eta) + d + \exp(-\eta)]^3$,

$E_{-3} = \lambda c^3 + cf^2,$

$E_{-2} = \omega cdf - k^2 cdf^2 + bf^2 - \omega bf^2 + k^2 bf^2 + 3\lambda bk^2 + 2cdf,$

$E_{-1} = 2cf + cd^2 + af^2 + 2bfd + 3\lambda b^2 c + 3\lambda ac^2 - 4k^2 cf + k^2 cd^2 - 2\omega af^2$
$\qquad + 2\omega cf + \omega cd^2 - \omega bfd - k^2 bfd + 4k^2 af^2,$

$E_0 = bd^2 + 2adf - 6k^2 bf - 3\omega adf + 3k^2 cd + 3\omega ad + 2cd + 2bf$
$\qquad + 3k^2 adf + 6\lambda abc + \lambda b^3,$

$E_1 = c + k^2 ad^2 - 4k^2 af + \omega db - \omega ad^2 - 2\omega af - k^2 db + 3\lambda a^2 c + 2bd + ad^2$
$\qquad + 4k^2 c + 3\lambda ab^2 + 2af + 2\omega c,$

$E_2 = 3\lambda a^2 b + \omega b + 2ad + k^2 b - k^2 ad - \omega ad + b,$

$E_3 = a + \lambda a^3.$

Equating the coefficients of $\exp(n\eta)$ in Eq. (10.13) to zero, we have the following system of equations:

$$E_{-3} = 0, \ E_{-2} = 0, \ E_{-1} = 0, \ E_0 = 0, \ E_1 = 0, \ E_2 = 0 \quad \text{and} \quad E_3 = 0. \tag{10.14}$$

By solving the above system of equations (10.14), we obtain

$$a = \pm\sqrt{\frac{-1}{\lambda}}, \ b = 0, \ c = 0, \ d = d, \ f = 0, \ \omega = \frac{3}{2} \text{ and } k = \pm\frac{\sqrt{2}}{2} \tag{10.15}$$

Case (i): For $\lambda = -\mu$ where $\mu > 0$ is an arbitrary constant, Eq. (10.4) becomes the *Newell–Whitehead equation* [11]. Then, we have

$$a = \pm\sqrt{\frac{1}{\mu}}, \ b = 0, \ c = 0, \ d = d, \ f = 0, \ \omega = \frac{3}{2} \text{ and } k = \pm\frac{\sqrt{2}}{2}.$$

The solution may be written from Eq. (10.12) as

$$u(x,t) = \frac{\pm\sqrt{\frac{1}{\mu}}\exp\left[\pm\frac{\sqrt{2}}{2}x + \frac{3}{2}t\right]}{\exp\left[\pm\frac{\sqrt{2}}{2}x + \frac{3}{2}t\right] + d} \tag{10.16}$$

where d is the free parameter.

Case (ii): For $\lambda = -1$, Eq. (10.4) is referred to as *Cahn–Allen equation* [11, 12]. Then the constants in Eq. (10.15) are $a = \pm 1, b = 0, c = 0, d = d, f = 0, \omega = \frac{3}{2}$, and $k = \pm\frac{\sqrt{2}}{2}$.

Accordingly, the solution of Eq. (10.4) for free parameter d and $\lambda = -1$ is

$$u(x,t) = \frac{\pm\exp\left[\pm\frac{\sqrt{2}}{2}x + \frac{3}{2}t\right]}{\exp\left[\pm\frac{\sqrt{2}}{2}x + \frac{3}{2}t\right] + d} = \frac{\pm 1}{1 + d\exp\left[\pm\frac{\sqrt{2}}{2}x + \frac{3}{2}t\right]} \tag{10.17}$$

The solution (10.16) is the same as the solution given in Refs. [11] and [12] for $d = \pm 1$.

Example 10.2 Let us now consider classical *Burger's equation* [13]

$$u_t = u_{xx} + \lambda u u_x. \tag{10.18}$$

Solution Equation (10.18) may be reduced to ordinary differential equation in η by using the above discussed transformation $u = u(\eta), \eta = kx + \omega t$ as

$$\omega u' - k^2 u'' - \lambda k u u' = 0 \tag{10.19}$$

where the prime denotes the differentiation with respect to η.

Next, we assume the solution of the ordinary differential equation (10.19) as Eq. (10.3). This may be rewritten as

$$u(\eta) = \frac{a_{-c}\exp(-c\eta) + \cdots + a_d\exp(d\eta)}{b_{-p}\exp(-p\eta) + \cdots + b_q\exp(q\eta)} \tag{10.20}$$

The highest-order linear and nonlinear terms in Eq. (10.19) are u'' and uu'. By differentiating and simplifying Eq. (10.20), we get

$$u'' = \frac{C_{AN}\exp[(-3p - c)\eta] + \cdots + C_{LN}\exp[(d + 3q)\eta]}{C_{AD}\exp[(-4p)\eta] + \cdots + C_{LD}\exp[(4q)\eta]} \tag{10.21}$$

and

$$uu' = \frac{D_{AN}\exp[(-p - 2c)\eta] + \cdots + D_{LN}\exp[(2d + q)\eta]}{D_{AD}\exp[(-3p)\eta] + \cdots + D_{LD}\exp[(3q)\eta]} \tag{10.22}$$

Again, as mentioned in Example 10.1, after differentiation and simplification we will get various terms containing different combinations of a_{-c}, a_d, b_{-p}, ..., etc. which for simplicity, we assumed as C_{AN}, ..., C_{LN}, C_{AD}, ..., C_{LD}, D_{AN}, ..., D_{LN}, and D_{AD}, ..., D_{LD}.

The lowest-order exp-function term of Eq. (10.8) is

$$\frac{D_{AN} \exp[(-p - 2c)\eta] + \cdots}{D_{AD} \exp[(-3p)\eta] + \cdots}.$$

Multiplying and dividing by $\exp(-p\eta)$, we get

$$\frac{D_{AN} \exp[(-2p - 2c)\eta] + \cdots}{D_{AD} \exp[(-4p)\eta] + \cdots} \tag{10.23}$$

Similarly, the highest order exp-function term of Eq. (10.8) is

$$\frac{\cdots + D_{LN} \exp[(2d + q)\eta]}{\cdots + D_{LD} \exp[(3q)]}.$$

Multiplying and dividing by $\exp(q\eta)$, we get

$$\frac{\cdots + D_{LN} \exp[(2d + 2q)\eta]}{\cdots + D_{LD} \exp[(4q)]}. \tag{10.24}$$

Next, we balance the lowest-order exp-functions in Eqs. (10.21) and (10.23) to have relation between p and c as below

$$-3p - c = -2p - 2c$$
$$\Rightarrow p = c.$$

Similarly, we balance the highest-order exp-functions in Eqs. (10.21) and (10.24) and we have

$$d + 3q = 2d + 2q$$
$$\Rightarrow d = q.$$

Here, for simplicity, we consider $p = c = 1$ and $d = q = 1$. Then the trial solution (10.3) takes the form

$$u(\eta) = \frac{a_{-1} \exp(-\eta) + a_0 + a_1 \exp(\eta)}{b_{-1} \exp(-\eta) + b_0 + b_1 \exp(\eta)}. \tag{10.25}$$

For the case $b_1 \neq 0$, Eq. (10.25) can be simplified as below

$$u(\eta) = \frac{a \exp(\eta) + b + c \exp(-\eta)}{\exp(\eta) + d + f \exp(-\eta)} \tag{10.26}$$

where a, b, c, d, and f are coefficients as explained in Example 10.1, which are to be determined next.

Plugging Eq. (10.26) in the ordinary differential equation (10.19) and simplifying, we get

$$\frac{1}{A}\left[E_{-2}\exp(-2\eta) + E_{-1}\exp(-\eta) + E_0 + E_1\exp(\eta) + E_2\exp(2\eta)\right] = 0 \quad (10.27)$$

where $A = [\exp(\eta) + d + \exp(-\eta)]^3$,

$$E_{-2} = -\lambda kcfb - \omega cdf + \lambda kc^2 d + \omega f^2 b - k^2 f^2 b + k^2 cdf,$$

$$E_{-1} = -2\lambda kcfa - \lambda kfb^2 + \lambda kbdc + k^2 dbf - k^2 cd^2 + 2\lambda kc^2 - 2\omega cf$$
$$\quad - 4k^2 af^2 - \omega cd^2 + 4k^2 cf + 2\omega af^2 + \omega dbf,$$

$$E_0 = 6k^2 bf - 3\omega cd - 3k^2 cd + 3\lambda kbc + 3\omega adf - 3k^2 adf - 3\lambda kafb,$$

$$E_1 = -\omega db + 2\lambda kac - 2\lambda ka^2 f - 2\omega c - 4k^2 c + k^2 db + \lambda kb^2 + \omega ad^2$$
$$\quad - k^2 ad^2 + 4k^2 af + 2\omega af - \lambda kbda,$$

$$E_2 = -\lambda ka^2 d - \omega b - k^2 b + \omega ad + \lambda kab + k^2 ad.$$

Equating the coefficients of $\exp(n\eta)$ in Eq. (10.27) to zero, we have

$$E_{-2} = 0, \quad E_{-1} = 0, \quad E_0 = 0, \quad E_1 = 0 \quad \text{and} \quad E_2 = 0. \quad (10.28)$$

By solving the above system of equations (10.28), we obtain $a = a$, $b = 0$, $c = \frac{f(-4k+\lambda a)}{\lambda}$, $d = 0, f = f, k = k$, and $\omega = \lambda ka - 2k^2$.

By substituting the above obtained values of b, c, d, and ω in Eq. (10.26), we get the final solution as

$$u(x,t) = \frac{1}{\lambda}\left[\frac{a\lambda\exp(kx + (\lambda ka - 2k^2)t) + f(-4k + \lambda a)\exp(-kx - (\lambda ka - 2k^2)t)}{\exp(kx + (\lambda ka - 2k^2)t) + f\exp(-kx - (\lambda ka - 2k^2)t)}\right]$$

where a, f, and k are free parameters which may be found from initial and boundary conditions.

Exercise

1 Solve the KdV equation $\frac{\partial u}{\partial t} + u\frac{\partial u}{\partial x} + \frac{\partial^3 u}{\partial x^3} = 0$ using the exp-function method.

2 Solve the modified KdV equation $\frac{\partial u}{\partial t} + u^2\frac{\partial u}{\partial x} + \frac{\partial^3 u}{\partial x^3} = 0$ using the exp-function method.

References

1 He, J.H. and Wu, X.H. (2006). Exp-function method for nonlinear wave equations. *Chaos, Solitons and Fractals* 30 (3): 700–708.
2 Dai, C.Q. and Zhang, J.F. (2009). Application of He's exp-function method to the stochastic mKdV equation. *International Journal of Nonlinear Sciences and Numerical Simulation* 10 (5): 675–680.

3 Misirli, E. and Gurefe, Y. (2010). Exact solutions of the Drinfel'd–Sokolov–Wilson equation using the exp-function method. *Applied Mathematics and Computation* 216 (9): 2623–2627.

4 Boz, A. and Bekir, A. (2008). Application of exp-function method for (3+1)-dimensional nonlinear evolution equations. *Computers and Mathematics with Applications* 56 (5): 1451–1456.

5 Bekir, A. (2010). Application of the exp-function method for nonlinear differential-difference equations. *Applied Mathematics and Computation* 215 (11): 4049–4053.

6 Zhang, S., Tong, J.L., and Wang, W. (2009). Exp-function method for a nonlinear ordinary differential equation and new exact solutions of the dispersive long wave equations. *Computers and Mathematics with Applications* 58 (11–12): 2294–2299.

7 He, J.H. and Abdou, M.A. (2007). New periodic solutions for nonlinear evolution equations using Exp-function method. *Chaos, Solitons and Fractals* 34 (5): 1421–1429.

8 Wu, X.H.B. and He, J.H. (2008). Exp-function method and its application to nonlinear equations. *Chaos, Solitons and Fractals* 38 (3): 903–910.

9 Bekir, A. and Boz, A. (2009). Application of He's exp-function method for nonlinear evolution equations. *Computers and Mathematics with Applications* 58 (11): 2286–2293.

10 Biazar, J. and Ayati, Z. (2009). Extension of the Exp-function method for systems of two-dimensional Burgers equations. *Computers and Mathematics with Applications* 58 (11): 2103–2106.

11 Wazwaz, A.M. (2007). The tanh–coth method for solitons and kink solutions for nonlinear parabolic equations. *Applied Mathematics and Computation* 188 (2): 1467–1475.

12 Taşcan, F. and Bekir, A. (2009). Travelling wave solutions of the Cahn–Allen equation by using first integral method. *Applied Mathematics and Computation* 207 (1): 279–282.

13 Yıldırım, A. and Pınar, Z. (2010). Application of the exp-function method for solving nonlinear reaction–diffusion equations arising in mathematical biology. *Computers and Mathematics with Applications* 60 (7): 1873–1880.

11

Adomian Decomposition Method

11.1 Introduction

The Adomian Decomposition Method (ADM) was first introduced by Adomian in the early 1980s [1, 2]. It is an efficient semi-analytical technique used for solving linear and nonlinear differential equations. It permits us to handle both nonlinear initial value problems (IVPs) and boundary value problems (BVPs).

The solution technique of this method [3, 4] is mainly based on decomposing the solution of nonlinear operator equation to a series of functions. Each presented term of the obtained series is developed from a polynomial generated in the expansion of an analytic function into a power series. Generally, the abstract formulation of this technique is very simple, but the actual difficulty arises while calculating the required polynomials and also in proving the convergence of the series of functions. In this chapter, we present procedures for solving linear as well as nonlinear ordinary/partial differential equations by the ADM along with example problems for clear understanding.

The following section addresses the ADM for handling ordinary differential equations (ODEs).

11.2 ADM for ODEs

The ADM mainly depends on decomposing the governing differential equation [5, 6] viz. $F(x, p(x), p'(x)) = 0$ into two components

$$L_x(p(x)) + N(p(x)) = 0 \qquad (11.1)$$

where $L_x = \frac{d}{dx}$ and N are the linear and nonlinear parts of F, respectively. The operator L_x is assumed as an invertible operator. Solving for $L_x(p(x))$, Eq. (11.1) leads to

$$L_x(p(x)) = -N(p(x)) \qquad (11.2)$$

Advanced Numerical and Semi-Analytical Methods for Differential Equations, First Edition.
Snehashish Chakraverty, Nisha Rani Mahato, Perumandla Karunakar, and Tharasi Dilleswar Rao.
© 2019 John Wiley & Sons, Inc. Published 2019 by John Wiley & Sons, Inc.

Applying the inverse operator $L_x^{-1} = \int_0^x dx$ on both sides of Eq. (11.2) yields

$$p(x) = \phi(x) - L_x^{-1}[N(p(x))] \tag{11.3}$$

where $p(0) = \phi(x)$ is the constant of integration which satisfies the given initial conditions. Now assume that the solution "$p(x)$" can be represented as an infinite series of the form

$$p(x) = \sum_{n=0}^{\infty} P_n(x) \tag{11.4}$$

Furthermore, from Eq. (11.4) the nonlinear term $N(p(x))$ is written as infinite series [1, 2] in terms of the Adomian polynomials $A_n(x)$ of the form

$$N\left(\sum_{n=0}^{\infty} P_n(x)\right) = \sum_{n=0}^{\infty} A_n(x) \tag{11.5}$$

where A_n Adomian polynomials of $p_0, p_1, p_2, \ldots, p_n$ are given by the following formula [3, 4]:

$$A_n(x) = \frac{1}{n!}\left[\frac{d^n}{dx^n}N\left(\sum_{i=0}^{\infty} p_i x^i\right)\right]_{x=0} \tag{11.6}$$

Then, substituting Eqs. (11.4) and (11.5) in Eq. (11.3) gives

$$\sum_{n=0}^{\infty} P_n(x) = \phi(x) - L^{-1}\left(\sum_{n=0}^{\infty} A_n(x)\right) \tag{11.7}$$

From Eq. (11.6), one may find the Adomian polynomials as below [4, 5]

$$
\left.\begin{aligned}
A_0(x) &= N(p_0), \\[2mm]
A_1(x) &= \frac{d}{dx}N(p_0 + xp_1)\bigg|_{x=0} = p_0 N'(p_0), \\[2mm]
A_2(x) &= \frac{1}{2!}\frac{d^2}{dx^2}N(p_0 + xp_1 + x^2 p_2)\bigg|_{x=0} = p_2(x)N'(p_0) + \frac{1}{2}p_1^2 N''(p_0) \\[2mm]
A_3(x) &= \frac{1}{3!}\frac{d^3}{dx^3}N\sum_{n=0}^{3} x^n p_n(x)\bigg|_{x=0} = p_3(x)N'(p_0) + p_1(x)p_3(x) + \frac{1}{3!}p_1^3 N'''(p_0) \\[2mm]
A_4(x) &= \frac{1}{4!}\frac{d^4}{dx^4}N\sum_{n=0}^{4} x^n p_n(x)\bigg|_{x=0} \\[2mm]
&= p_4(x)N'(p_0) + \left(p_1(x)p_3(x) + \frac{1}{2}p_1^2(x)\right)N''(p_0) + \frac{1}{2}p_1^2(x)p_2(x)N'''(p_0) \\[2mm]
&\quad + \frac{1}{4!}p_1^4 N''''(p_0) \\[2mm]
&\vdots
\end{aligned}\right\} \tag{11.8}
$$

In order to get the solution $p(x) = \sum_{n=0}^{\infty} p_n(x)$, the process is longer but in practice all the terms of the series may not be required and the solution may be approximated by the truncated series $\sum_{n=0}^{N} p_n$ (by using the convergence of the series).

Next, we present linear and nonlinear IVPs for clear understanding of the ADM for ODEs.

Example 11.1   Consider the IVP $e^x y'' + xy = 0$ subject to the initial conditions $y(0) = 2, y'(0) = 5$.

Solution   To apply the ADM, we rewrite the given equation in an operator form as

$$L_{xx}y = -xe^{-x}y \tag{11.9}$$

where $L_{xx} = \frac{d^2}{dx^2}$ and $L_{xx}^{-1} = \int_0^x \int_0^s dx ds$. By comparing Eq. (11.9) with Eq. (11.1) the nonlinear part $N(p(x)) = 0$.

By applying L_{xx}^{-1} to both sides of Eq. (11.9), we get

$$y(x) = 2 + 5x - \int_0^x \int_0^s xe^{-x}y(x)dxds. \tag{11.10}$$

where $\phi(x) = 2 + 5x$ by comparing Eq. (11.10) with Eq. (11.3) which satisfies the given initial conditions.

Now assume that the solution "$y(x)$" can be represented as infinite series of the form

$$y(x) = \sum_{n=0}^{\infty} y_n(x). \tag{11.11}$$

By substituting Eq. (11.11) in both sides of Eq. (11.10) yields

$$\sum_{n=0}^{\infty} y_n(x) = 2 + 5x - \int_0^x \left[\int_0^s \left[xe^{-x} \sum_{n=0}^{\infty} y_n(x) dx \right] \right] ds. \tag{11.12}$$

Next, by using Eqs. (11.5), (11.8), and (11.12), one can obtain the Adomian polynomials $\sum_{n=0}^{\infty} A_n(x)dx = \sum_{n=0}^{\infty} y_n(x)dx$ (because, as mentioned above there is no nonlinear part in the considered problem) as

$$y_0(x) = 2 + 5x,$$

$$y_1(x) = -L_{xx}^{-1}(xe^{-x}y_0(x)),$$

$$= -L_{xx}^{-1}(xe^{-x}(2 + 5x)),$$

$$= -L_{xx}^{-1}(2xe^{-x} + 5x^2e^{-x}),$$

$$\vdots$$

$$\sum_{n=0}^{\infty} y_n(x) = -L_{xx}^{-1}\left(2\sum_{n=0}^{\infty}\frac{(-1)^n x^{n+1}}{n!} + 5\sum_{n=0}^{\infty}\frac{(-1)^n x^{n+2}}{n!}\right),$$

$$\Rightarrow \sum_{n=0}^{\infty} y_n(x) = \left(2\sum_{n=0}^{\infty}\frac{(-1)^n x^{n+3}}{(n+3)(n+2)n!} - 5\sum_{n=0}^{\infty}\frac{(-1)^n x^{n+4}}{(n+4)(n+3)n!}\right).$$

$$(11.13)$$

where $e^{-x} = \frac{(-1)^n x^n}{n!}$.

From Eqs. (11.13) and (11.11), we can obtain

$$y(x) = 2\left(\frac{1}{6}x^3 - \frac{1}{212}x^4 + \frac{1}{40}x^3 + \cdots\right) + 5\left(\frac{1}{12}x^4 - \frac{1}{20}x^5 + \frac{1}{60}x^6 + \cdots\right)$$

Example 11.2 Consider the nonlinear IVP $\frac{du}{dx} = \frac{u^2}{1-xu}$, $u(0) = 1$.

Solution The nonlinear IVP $\frac{du}{dx} = \frac{u^2}{1-xu}$ can be reduced to the operator form

$$L_x u = xuu' + u^2, \quad u(0) = 1. \tag{11.14}$$

In this case, $L_x = \frac{d}{dx}$ is the ordinary differential operator and $L_x^{-1} = \int_0^x dx$ is an integral operator. By applying L_x^{-1} to both sides of Eq. (11.14), we get

$$u(x) = u(0) + L_x^{-1}(xuu') + L_x^{-1}(u^2). \tag{11.15}$$

By using the initial condition, we obtain,

$$u(x) = 1 + L_x^{-1}(xuu') + L_x^{-1}(u^2). \tag{11.16}$$

Then, in the decomposition method, each of the nonlinear terms uu' and u^2 are formally expanded in terms of power series given by [4, 5]

$$uu' = \sum_{n=0}^{\infty} A_n; \quad u^2 = \sum_{n=0}^{\infty} B_n \tag{11.17}$$

where A_n and B_n are the Adomian polynomials corresponding to the nonlinear terms uu' and u^2, respectively. The Adomian polynomials are further given (Eq. (11.8)) by

$$\left.\begin{aligned}
A_0 &= u_0 u'_0, \\
A_1 &= u'_0 u_1 + u'_1 u_0, \\
A_2 &= u'_0 u_2 + u'_1 u_1 + u'_2 u_0, \\
&\vdots
\end{aligned}\right\} \tag{11.18}$$

and

$$
\left.\begin{aligned}
B_0 &= u_0^2, \\
B_1 &= 2u_0' u_1, \\
B_2 &= 2u_0' u_2 + u_1^2, \\
&\vdots
\end{aligned}\right\}
\tag{11.19}
$$

By substituting Eq. (11.17) in Eq. (11.16), we have

$$
\sum_{n=0}^{\infty} u_n = 1 + L_x^{-1}\left(\sum_{n=0}^{\infty} x A_n\right) + L_x^{-1}\left(\sum_{n=0}^{\infty} B_n\right).
\tag{11.20}
$$

On comparing both sides of the series of Eq. (11.20), we obtain $u_{k+1}(x) = L_x^{-1}(x A_K) + L_x^{-1}(B_K)$, $k \geq 0$. By using Eqs. (11.18) and (11.19), this leads to

$$
u_0(x) = 1,
$$

$$
u_1(x) = L_x^{-1}(x A_0) + L_x^{-1}(B_0) = x,
$$

$$
u_2(x) = L_x^{-1}(x A_1) + L_x^{-1}(B_1) = \frac{3}{2}x^2,
$$

$$
u_3(x) = L_x^{-1}(x A_2) + L_x^{-1}(B_2) = \frac{8}{3}x^3,
$$

$$
u_4(x) = L_x^{-1}(x A_4) + L_x^{-1}(B_4) = \frac{125}{4}x^4, \ldots,
$$

The solution of Eq. (11.14) $u(x) = \sum_{n=0}^{\infty} u_n$ may now be written as

$$
u(x) = 1 + x + \frac{3}{2}x^2 + \frac{8}{3}x^3 + \frac{125}{4}x^4 + \cdots
$$

11.3 Solving System of ODEs by ADM

A system of first-order ordinary differential equations can be considered as [7, 8],

$$
\left.\begin{aligned}
u_1' &= f_1(x, u_1, u_2, \ldots, u_n) \\
u_2' &= f_2(x, u_1, u_2, \ldots, u_n) \\
&\vdots \\
u_n' &= f_n(x, u_1, u_2, \ldots, u_n)
\end{aligned}\right\}
\tag{11.21}
$$

We can present the system of equations (11.21) as

$$L_x u_i = f_i(x, u_1, u_2, \dots, u_n) \quad \text{for} \quad i = 1, 2, \dots, n. \tag{11.22}$$

where L_x is the linear operator $\frac{d}{dx}$ with inverse linear operator, $L_x^{-1} = \int_0^x dx$. Applying the inverse operator on Eq. (11.22), we get the following canonical form which is suitable for applying the ADM.

$$u_i(x) = u_i(0) + \int_0^x f_i(x, u_1, u_2, \dots, u_n)dx, \quad i = 1, 2, \dots, n \tag{11.23}$$

where $u_i(0)$ are the initial conditions of Eq. (11.22).

As mentioned earlier, the solution of Eq. (11.22) using the ADM is considered as the sum of a series

$$u_i = \sum_{j=0}^{\infty} f_{i,j} \tag{11.24}$$

and the integrand in Eq. (11.23) reduces to

$$f_i(x, u_1, \dots, u_n) = \sum_{j=0}^{\infty} A_{i,j}(f_{i,0}, f_{i,1}, \dots, f_{i,j}) \tag{11.25}$$

where $A_{i,j}(f_{i,0}, f_{i,1}, \dots, f_{i,n})$ are the Adomian polynomials. Substituting Eqs. (11.24) and (11.25) in Eq. (11.23), we get

$$\sum_{j=0}^{\infty} f_{i,j} = u_i(0) + \sum_{j=0}^{\infty} \int_0^x A_{i,j}(f_{i,0}, f_{i,1}, \dots, f_{i,j})dx, \tag{11.26}$$

From Eq. (11.26), we have the following recursive relations as

$$f_{i,0} = u_i(0),$$
$$\vdots$$
$$f_{i,n+1} = \int_0^x A_{i,n}(f_{i,0}, f_{i,1}, \dots f_{i,n})dx, \quad n = 0, 1, 2, \dots \tag{11.27}$$

By solving Eq. (11.26), we obtain $f_{i,j}$ and $u_i = \sum_{j=0}^{\infty} f_{i,j}$.

Example 11.3 Let us consider the following system of differential equations with initial conditions $u_1(0) = 1$, $u_2(0) = 0$, and $u_3(0) = 2$,

$$\left. \begin{array}{l} u_1' = u_3 - \cos x, \\ u_2' = u_3 - e^x, \\ u_3' = u_1 - u_2. \end{array} \right\} \tag{11.28a}$$

By applying the inverse operator $L^{-1} = \int_0^x dx$ to both sides of Eq. (11.28a), we get

$$
\left.
\begin{aligned}
u_1 &= 1 - \int_0^x \cos x \, dx + \int_0^x u_3 dx \\
u_2 &= - \int_0^x e^x dx + \int_0^x u_3 dx \\
u_3 &= 2 + \int_0^x (u_1 - u_2) dx.
\end{aligned}
\right\}
\tag{11.28b}
$$

General form of the system (11.28b) for computing Adomian polynomials (11.27) can be written as

$$
\left.
\begin{aligned}
u_{1,0} &= 1 - \sin x, & u_{1,n+1} &= L^{-1} u_{3,n} \\
u_{2,0} &= 1 - e^x, & u_{2,n+1} &= L^{-1} u_{3,n} \\
u_{3,0} &= 2, & u_{3,n+1} &= L^{-1}(u_{1,n} - u_{2,n}).
\end{aligned}
\right\}
\tag{11.28c}
$$

where $n = 0, 1, 2, \dots$.

By expanding and solving the system of equations (11.28c), we get the solution of Eq. (11.28a) as $u_1 = e^x$, $u_2 = \sin x$, and $u_3 = e^x + \cos x$.

11.4 ADM for Solving Partial Differential Equations

Let us consider a general nonlinear partial differential equations (PDEs) in the following form [9]:

$$
Lu + Ru + Nu = g
\tag{11.29}
$$

The linear term of nonlinear PDE is represented as $Lu + Ru$, while the nonlinear term is represented by Nu. Here, L is the invertible linear operator and R is the remaining linear part. By applying inverse operator $L^{-1} = \int_0^x dx$, Eq. (11.29) can be written as

$$
u = \phi + L^{-1}(g) - L^{-1}(Ru) - L^{-1}(Nu)
\tag{11.30}
$$

where $\phi(x)$ is the constant of integration which satisfies the given initial conditions.

Then, the solution may be represented by the decomposition method given in Eq. (11.30) by the following infinite series:

$$
u = \sum_{n=0}^{\infty} u_n
\tag{11.31}
$$

The nonlinear operator Nu is then decomposed as

$$Nu = \sum_{n=0}^{\infty} A_n. \tag{11.32}$$

where $A_n's$ are Adomian polynomials defined as

$$A_n = \frac{1}{n!} \frac{d^n}{dx^n} \left[N \left(\sum_{n=1}^{n} x^i u_i \right) \right] \Bigg|_{x=0}, \quad n = 0, 1, 2, \ldots \tag{11.33}$$

By substituting Eqs. (11.31) and (11.32) in Eq. (11.30), we have

$$u = \sum_{n=0}^{\infty} u_n = \phi + L^{-1}(g) - L^{-1} \left(R \left(\sum_{n=0}^{\infty} u_n \right) \right) - L^{-1} \left(\sum_{n=0}^{\infty} A_n \right) \tag{11.34}$$

Consequently, Eq. (11.34) can be written in terms of recursive relations as

$$\left. \begin{aligned} u_0 &= \phi + L^{-1}(g), \\ u_1 &= -L^{-1}(R(u_0)) - L^{-1}(A_0) \\ u_2 &= -L^{-1}(R(u_1)) - L^{-1}(A_1) \\ &\vdots \\ u_n &= -L^{-1}(R(u_{n-1})) - L^{-1}(A_{n-1}). \end{aligned} \right\} \tag{11.35}$$

Hence, all the terms of "u" may be calculated using the above recursive relations and the general solution obtained according to the ADM $u = \sum_{n=0}^{\infty} u_n$.

Example 11.4   Consider the following hyperbolic nonlinear problem

$$\frac{\partial u}{\partial t} = u \frac{\partial u}{\partial x}, \quad 0 \le x \le 1, \quad 0 \le t \le 1, \tag{11.36}$$

with the initial condition $u(x, 0) = \frac{x}{10}$.

Solution   By using the ADM we have $Ru = 0$, $Lu = \frac{\partial u}{\partial t}$, $Nu = u \frac{\partial u}{\partial x}$, $g(x, t) = 0$, and $\phi = u(x, 0) = \frac{x}{10}$.

The approximate solution of Eq. (11.36) may be represented as

$$u = \sum_{n=0}^{\infty} u_n. \tag{11.37}$$

By using Eq. (11.35) the Adomian polynomials are obtained as

$$u_0 = \frac{x}{10}, \quad u_1 = \frac{x}{10}\left(\frac{t}{10}\right), \quad u_2 = \frac{x}{10}\left(\frac{t}{10}\right)^2,$$

$$u_3 = \frac{x}{10}\left(\frac{t}{10}\right)^3, \dots, u_n = \frac{x}{10}\left(\frac{t}{10}\right)^n$$

By substituting these individual terms in Eq. (11.37), we obtain

$$u(x, t) = \frac{x}{10}\left[1 + \frac{t}{10} + \left(\frac{t}{10}\right)^2 + \left(\frac{t}{10}\right)^3 + \cdots + \left(\frac{t}{10}\right)^n + \cdots\right].$$

11.5 ADM for System of PDEs

ADM transforms system of PDEs into a set of recursive relation that can easily be handled [9, 10]. To understand the method, we now consider the following system of linear PDEs:

$$\left.\begin{array}{l} \dfrac{\partial p}{\partial t} + \dfrac{\partial q}{\partial x} = f_1 \\[2mm] \dfrac{\partial q}{\partial t} + \dfrac{\partial p}{\partial x} = f_2 \end{array}\right\} \tag{11.38}$$

subject to the initial conditions, $p(x, 0) = g_1(x)$; $q(x, 0) = g_2(x)$.

The system of linear PDEs, given in Eq. (11.38), can be further written as

$$\left.\begin{array}{l} L_t p + L_x q = f_1 \\ L_t q + L_x p = f_2 \end{array}\right\} \tag{11.39}$$

where L_t and L_x are first-order partial operators, and f_1 and f_2 are homogeneous terms. Now, by applying the inverse operator L_t^{-1} to Eq. (11.39) and then using the initial conditions we get

$$p(x, t) = g_1(x) + L_t^{-1} f_1 - L_t^{-1} L_x q, \tag{11.40}$$

$$q(x, t) = g_2(x) + L_t^{-1} f_2 - L_t^{-1} L_x p. \tag{11.41}$$

The ADM suggests that the linear terms $p(x, t)$ and $q(x, t)$ are decomposed by infinite series,

$$p(x, t) = \sum_{n=0}^{\infty} p_n(x, t). \tag{11.42}$$

and

$$q(x, t) = \sum_{n=0}^{\infty} q_n(x, t). \tag{11.43}$$

By substituting Eqs. (11.42) and (11.43) in Eqs. (11.40) and (11.41), respectively, we obtain

$$\sum_{n=0}^{\infty} p_n(x, t) = g_1(x) + L_t^{-1}f_1 - L_t^{-1}L_x \sum_{n=0}^{\infty} q_n(x, t). \tag{11.44}$$

and

$$\sum_{n=0}^{\infty} q_n(x, t) = g_2(x) + L_t^{-1}f_2 - L_t^{-1}L_x \sum_{n=0}^{\infty} p_n(x, t). \tag{11.45}$$

The system is then transformed into a set of following recursive relations:

$$p_0(x, t) = g_1(x) + L_t^{-1}f_1,$$
$$p_{k+1}(x, t) = -L_t^{-1}L_x q_k(x, t), \quad k \geq 0$$

and

$$q_0(x, t) = g_2(x) + L_t^{-1}f_2,$$
$$q_{k+1}(x, t) = -L_t^{-1}L_x p_k(x, t), \quad k \geq 0$$

Example 11.5   Solve the following linear system of PDEs:

$$p_t + q_x = 0,$$
$$q_t + p_x = 0.$$

with the initial conditions $p(x, 0) = e^x$; $q(x, 0) = e^{-x}$.

Solution   The above system of linear PDEs can be written in terms of differential operator as

$$\left. \begin{array}{l} L_t p + L_x q = 0 \\ L_t q + L_x p = 0 \end{array} \right\} \tag{11.46}$$

subject to the initial conditions $p(x, 0) = e^x$, $q(x, 0) = e^{-x}$.

In order to obtain the solution by using the ADM, we obtain the recursive relations,

$$p_0(x, t) = e^x$$
$$p_{k+1}(x, t) = -L_t^{-1}L_x q_k(x, t), \quad k \geq 0$$

and

$$q_0(x, t) = e^{-x}$$
$$q_{k+1}(x, t) = -L_t^{-1}L_x p_k(x, t), \quad k \geq 0$$

The remaining components are thus determined by

$$
\left.
\begin{aligned}
(p_0, q_0) &= (e^x, e^{-x}) \\
(p_1, q_1) &= (te^{-x}, -te^x) \\
(p_2, q_2) &= \left(\frac{t^2}{2!} e^x, \frac{t^2}{2!} e^{-x} \right) \\
(p_3, q_3) &= \left(\frac{t^3}{3!} e^{-x}, -\frac{t^3}{3!} e^x \right) \\
&\;\;\vdots
\end{aligned}
\right\}
\tag{11.47}
$$

By using Eq. (11.47), the unknowns p_i and q_i for $i = 1, 2, \dots$ are computed and substituted in Eqs. (11.42) and (11.43) to obtain the solution as

$$
p(x, t) = p_0 + p_1 + p_2 + \cdots = e^x \left(1 + \frac{t^2}{2!} + \frac{t^4}{4!} + \cdots \right)
$$
$$
+ e^{-x} \left(t + \frac{t^3}{3!} + \frac{t^5}{5!} + \cdots \right) = e^x \cosh t + e^{-x} \sinh t,
$$
$$
q(x, t) = q_0 + q_1 + q_2 + \cdots = e^{-x} \left(1 + \frac{t^2}{2!} + \frac{t^4}{4!} + \cdots \right)
$$
$$
- e^x \left(t + \frac{t^3}{3!} + \frac{t^5}{5!} + \cdots \right) = e^{-x} \cosh t - e^x \sinh t.
$$

Exercise

1 Solve the following IVPs by using the ADM

$$
\frac{dy}{dx} = \frac{y^2}{1 - xy}, \quad y(0) = 1
$$
$$
\frac{dy}{dx} = \frac{y}{x - y}, \quad y(0) = 1
$$

2 Solve the linear system of PDEs

$$
p_t + q_x + 2q = 0
$$
$$
q_t + p_x + 2p = 0
$$

with the initial data

$$
p(x, 0) = \cos x; \quad q(x, 0) = \sin x
$$

3 Solve the problem

$$\frac{\partial y}{\partial t} = x^2 - \frac{1}{4}\left(\frac{\partial y}{\partial x}\right)^2$$

with the initial condition

$$y(x, 0) = 0, \quad 0 \le x \le 1$$

References

1 Wazwaz, A.M. (1998). A comparison between Adomian decomposition method and Taylor series method in the series solutions. *Applied Mathematics and Computation* 97 (1): 37–44.

2 Adomian, G. (1990). A review of the decomposition method and some recent results for nonlinear equations. *Mathematical and Computer Modelling* 13 (7): 17–43.

3 Momani, S. and Odibat, Z. (2006). Analytical solution of a time-fractional Navier–Stokes equation by Adomian decomposition method. *Applied Mathematics and Computation* 177 (2): 488–494.

4 Evans, D.J. and Raslan, K.R. (2005). The Adomian decomposition method for solving delay differential equation. *International Journal of Computer Mathematics* 82 (1): 49–54.

5 Jafari, H. and Daftardar-Gejji, V. (2006). Revised Adomian decomposition method for solving systems of ordinary and fractional differential equations. *Applied Mathematics and Computation* 181 (1): 598–608.

6 Batiha, B., Noorani, M.S.M., and Hashim, I. (2008). Numerical solutions of the nonlinear integro-differential equations. *International Journal of Open Problems in Computer Science* 1 (1): 34–42.

7 Dehghan, M., Shakourifar, M., and Hamidi, A. (2009). The solution of linear and nonlinear systems of Volterra functional equations using Adomian–Pade technique. *Chaos, Solitons, and Fractals* 39 (5): 2509–2521.

8 Jafari, H. and Daftardar-Gejji, V. (2006). Revised Adomian decomposition method for solving a system of nonlinear equations. *Applied Mathematics and Computation* 175 (1): 1–7.

9 Bildik, N. and Konuralp, A. (2006). The use of variational iteration method, differential transform method and Adomian decomposition method for solving different types of nonlinear partial differential equations. *International Journal of Nonlinear Sciences and Numerical Simulation* 7 (1): 65–70.

10 Wazwaz, A.M. (1999). A reliable modification of Adomian decomposition method. *Applied Mathematics and Computation* 102 (1): 77–86.

12

Homotopy Perturbation Method

12.1 Introduction

Homotopy perturbation method (HPM) is a semi-analytical technique for solving linear as well as nonlinear ordinary/partial differential equations. The method may also be used to solve a system of coupled linear and nonlinear differential equations. The HPM was proposed by J. He in 1999 [1]. This method was developed by making use of artificial parameters [2]. Interested readers may go through Refs. [3–6] for further details.

Almost all traditional perturbation methods are based on small parameter assumption. But, a majority of nonlinear problems have no small parameters at all and the determination of small parameters seems to be a special art requiring special techniques. These small parameters are so sensitive, such that a small change in small parameters will affect the results. An appropriate choice of small parameters leads to ideal results. However, an unsuitable choice of small parameters results in bad effects, sometimes seriously. Liu [2] proposed artificial parameter method and Liao [7, 8] contributed homotopy analysis method to eliminate small parameter assumption. Further, He [1, 9] developed two effective techniques viz. variational iteration method (VIM) and HPM in which no small parameter assumptions are required, where details of the VIM are given in Chapter 13.

12.2 Basic Idea of HPM

In this section we illustrate the basic idea of the HPM. For this we consider the following differential equation:

$$A(u) - f(r) = 0, \quad r \in \Omega \tag{12.1}$$

subject to the boundary condition

$$B\left(u, \frac{\partial u}{\partial r}\right) = 0, \quad r \in \Gamma \tag{12.2}$$

Advanced Numerical and Semi-Analytical Methods for Differential Equations, First Edition.
Snehashish Chakraverty, Nisha Rani Mahato, Perumandla Karunakar, and Tharasi Dilleswar Rao.
© 2019 John Wiley & Sons, Inc. Published 2019 by John Wiley & Sons, Inc.

where A represents a general differential operator, B is a boundary operator, Γ is the boundary of the domain Ω, and $f(r)$ is known analytic function.

The operator A can be decomposed into two parts viz. linear (L) and nonlinear (N). Therefore, Eq. (12.1) may be written in the following form:

$$L(u) + N(u) - f(r) = 0 \tag{12.3}$$

An artificial parameter p can be embedded in Eq. (12.3) as follows:

$$L(u) + p(N(u) - f(r)) = 0, \tag{12.4}$$

where $p \in [0, 1]$ is the embedding parameter (also called as an artificial parameter).

Using homotopy technique, proposed by He [1] and Liao [7, 8], we construct a homotopy $v(r, p) : \Omega \times [0, 1] \to R$ to Eq. (12.3) which satisfies

$$H(v, p) = (1 - p)[L(v) - L(u_0)] + p[L(v) + N(v) - f(r)] = 0 \tag{12.5}$$

and

$$H(v, p) = L(v) - L(u_0) + pL(u_0) + p[N(v) - f(r)] = 0. \tag{12.6}$$

Here, u_0 is an initial approximation of Eq. (12.6) which satisfies the given conditions.

By substituting $p = 0$ and $p = 1$ in Eq. (12.6), we may get the following equations, respectively.

$$H(v, 0) = L(v) - L(u_0)$$

and

$$H(v, 1) = A(v) - f(r) = 0.$$

As p changes from zero to unity, $v(r, p)$ changes from $u_0(r)$ to $u(r)$. In topology, this is called deformation and $L(v) - L(u_0)$ and $A(v) - f(r)$ are homotopic to each other. Due to the fact that $p \in [0, 1]$ is a small parameter, we consider the solution of Eq. (12.5) as a power series in p as below

$$v = v_0 + p v_1 + p^2 v_2 + \cdots \tag{12.7}$$

The approximate solution of Eq. (12.1) may then be obtained as

$$u = \lim_{p \to 1} v = v_0 + v_1 + v_2 + \cdots \tag{12.8}$$

The convergence of the series solution (12.8) has been given in Ref. [1].

12.3   Numerical Examples

Here, we use the present method to solve a nonlinear ordinary differential equation in Example 12.1 and a nonlinear partial differential equation in Example 12.2. Finally, a coupled nonlinear system of partial differential equations has been solved in Example 12.3. It is worth mentioning that the HPM can also be used for handling linear ordinary and partial differential equations.

Example 12.1   We consider a homogeneous ordinary differential equation from Ref. [5] as

$$\frac{dy}{dx} + y^2 = 0 \tag{12.9}$$

subject to the initial condition $y(0) = 1$.

Solution   By using the HPM, we construct homotopy $v(r, p) : \Omega \times [0, 1] \to R$ which satisfies

$$(1 - p)\left(\frac{dv}{dx} - \frac{dy_0}{dx}\right) + p\left(\frac{dv}{dx} + v^2\right) = 0, \tag{12.10}$$

$$\frac{\partial v}{\partial x} = \frac{\partial y_0}{\partial x} - p\frac{\partial y_0}{\partial x} - pv^2. \tag{12.11}$$

Solution of Eq. (12.10) is assumed in the form of a power series

$$v = v_0 + pv_1 + p^2 v_2 + \cdots \tag{12.12}$$

where $y = \lim_{p \to 1} v = v_0 + v_1 + v_2 + \cdots$ using Eq. (12.8).

By substituting Eq. (12.12) in Eq. (12.11), we get

$$\frac{\partial}{\partial x}(v_0 + pv_1 + p^2 v_2 + \cdots) = \frac{\partial y_0}{\partial x} - p\frac{\partial y_0}{\partial x} - p(v_0 + pv_1 + p^2 v_2 + \cdots)^2 \tag{12.13}$$

By equating like powers of p from both sides in Eq. (12.13), one may get

$$p^0 : \frac{dv_0}{dx} = \frac{dy_0}{dx} \tag{12.14}$$

$$p^1 : \frac{dv_1}{dx} = -\frac{dy_0}{dx} - v_0^2 \tag{12.15}$$

$$p^2 : \frac{dv_2}{dx} = -2v_0 v_1 \tag{12.16}$$

$$p^3 : \frac{dv_3}{dx} = -2v_0 v_2 - v_1^2 \tag{12.17}$$

$$p^4 : \frac{dv_4}{dx} = -2v_0v_3 - 2v_1v_2 \tag{12.18}$$

and so on.

By integrating above Eqs. (12.14)–(12.18) with respect to "x" we get

$$v_0 = y_0 = 1$$
$$v_1 = -x$$
$$v_2 = x^2.$$
$$v_3 = -x^3$$
$$v_4 = x^4.$$

The solution of Eq. (12.9) is

$$y = \lim_{p \to 1}(v_0 + pv_1 + p^2v_2 + \cdots)$$
$$\Rightarrow y = 1 - x + x^2 - x^3 + x^4 + \cdots$$
$$\Rightarrow y = \frac{1}{1+x}$$

Example 12.2 We consider a nonhomogeneous advection equation from Ref. [10] as

$$u_t + uu_x = x \tag{12.19}$$

subject to the initial condition $u(x, 0) = 2$.

Solution In the present problem, $L(u) = u_t$ is the linear part and $N(u) = uu_x$ is the nonlinear part of the nonlinear differential Eq. (12.19), and the known analytic function is $f(x) = x$.

Using the HPM as discussed above, we construct homotopy $v(r, p) : \Omega \times [0, 1] \to R$ which satisfies

$$(1-p)\left(\frac{\partial v}{\partial t} - \frac{\partial u_0}{\partial t}\right) + p\left(\frac{\partial v}{\partial t} + v\frac{\partial v}{\partial x} - x\right) = 0 \tag{12.20}$$

This after simplification becomes

$$\frac{\partial v}{\partial t} = \frac{\partial u_0}{\partial t} - p\frac{\partial u_0}{\partial t} - p\left(v\frac{\partial v}{\partial x} - x\right). \tag{12.21}$$

Assume the solution of Eq. (12.20) in the form of a power series as below

$$v = v_0 + pv_1 + p^2v_2 + \cdots \tag{12.22}$$

where $u = \lim_{p \to 1} v = v_0 + v_1 + v_2 + \cdots$ using Eq. (12.8).

By substituting Eq. (12.22) in Eq. (12.21), we have

$$\frac{\partial}{\partial t}(v_0 + pv_1 + p^2 v_2 + \cdots) = \frac{\partial u_0}{\partial t} - p\frac{\partial u_0}{\partial t}$$

$$- p\left((v_0 + pv_1 + p^2 v_2 + \cdots)\frac{\partial}{\partial x}(v_0 + pv_1 + p^2 v_2 + \cdots) - x\right) \quad (12.23)$$

By equating like powers of p from both sides of Eq. (12.23), we get

$$p^0 : \frac{\partial v_0}{\partial t} = \frac{\partial u_0}{\partial t} \quad (12.24)$$

$$p^1 : \frac{\partial v_1}{\partial t} = -\frac{\partial u_0}{\partial t} - v_0\frac{\partial v_0}{\partial x} + x \quad (12.25)$$

$$p^2 : \frac{\partial v_2}{\partial t} = -v_0\frac{\partial v_1}{\partial x} - v_1\frac{\partial v_0}{\partial x} \quad (12.26)$$

$$p^3 : \frac{\partial v_3}{\partial t} = -v_0\frac{\partial v_2}{\partial x} - v_1\frac{\partial v_1}{\partial x} - v_2\frac{\partial v_0}{\partial x} \quad (12.27)$$

$$p^4 : \frac{\partial v_4}{\partial t} = -v_0\frac{\partial v_3}{\partial x} - v_1\frac{\partial v_2}{\partial x} - v_2\frac{\partial v_1}{\partial x} - v_3\frac{\partial v_0}{\partial x} \quad (12.28)$$

$$p^5 : \frac{\partial v_5}{\partial t} = -v_0\frac{\partial v_4}{\partial x} - v_1\frac{\partial v_3}{\partial x} - v_2\frac{\partial v_2}{\partial x} - v_3\frac{\partial v_1}{\partial x} - v_4\frac{\partial v_0}{\partial x} \quad (12.29)$$

and so on.

The general term may be written as

$$p^{n+1} : \frac{\partial v_{n+1}}{\partial t} = -\sum_{i=0}^{n} v_i\frac{\partial v_{n-i}}{\partial x}, \quad n = 1, 2, 3, \ldots$$

By integrating above Eqs. (12.24)–(12.29) with respect to "t", one may obtain

$$v_0 = u_0 = 2$$
$$v_1 = xt$$
$$v_2 = -t^2$$
$$v_3 = -\frac{1}{3}xt^3$$
$$v_4 = \frac{5}{12}t^4$$
$$v_5 = \frac{2}{15}xt^5$$
$$\vdots$$

The solution of the advection equation as $p \to 1$ (using Eq. (12.8)) may then be written as

$$u = 2 + xt - t^2 - \frac{1}{3}xt^3 + \frac{5}{12}t^4 + \frac{2}{15}xt^5 + \cdots$$

$$u = 2\left(1 - \frac{1}{2}t^2 + \frac{5}{24}t^4 - \cdots\right) + x\left(t - \frac{1}{3}t^3 + \frac{2}{15}t^5 - \cdots\right)$$

$$u = 2\operatorname{sech} t + x \tanh t.$$

Example 12.3 Next, we apply the HPM to solve a coupled nonlinear system of Burgers' equations [11]:

$$u_t - u_{xx} - 2uu_x + (uv)_x = 0, \tag{12.30}$$

$$v_t - v_{xx} - 2vv_x + (uv)_x = 0, \tag{12.31}$$

subject to the initial conditions:

$$u(x,0) = \sin x \quad \text{and} \quad v(x,0) = \sin x. \tag{12.32}$$

Solution By using modified HPM proposed by Sweilam and Khader [11], we construct homotopies as follows:

$$u_t = u_t(x,0) + p[u_{xx} + 2uu_x - (uv)_x], \tag{12.33}$$

$$v_t = v_t(x,0) + p[v_{xx} + 2vv_x - (uv)_x], \tag{12.34}$$

We consider solutions of (12.30) and (12.31) in the series form as

$$u = u_0 + pu_1 + p^2 u_2 + \cdots \tag{12.35}$$

$$v = v_0 + pv_1 + p^2 v_2 + \cdots \tag{12.36}$$

By substituting Eqs. (12.35) and (12.36) in Eqs. (12.33) and (12.34), respectively, we get

$$\left. \begin{aligned} &\frac{\partial}{\partial t}(u_0 + pu_1 + p^2 u_2 + \cdots) \\ &= \frac{\partial}{\partial t}u(x,0) + p\Big[\frac{\partial^2}{\partial x^2}(u_0 + pu_1 + p^2 u_2 + \cdots) \\ &\quad + 2(u_0 + pu_1 + p^2 u_2 + \cdots)\frac{\partial}{\partial x}(u_0 + pu_1 + p^2 u_2 + \cdots) \\ &\quad - \frac{\partial}{\partial x}((u_0 + pu_1 + p^2 u_2 + \cdots)(v_0 + pv_1 + p^2 v_2 + \cdots))\Big], \end{aligned} \right\} \tag{12.37}$$

$$\left. \begin{aligned} &\frac{\partial}{\partial t}(v_0 + pv_1 + p^2 v_2 + \cdots) \\ &= \frac{\partial}{\partial t}v(x,0) + p\Big[\frac{\partial^2}{\partial x^2}(v_0 + pv_1 + p^2 v_2 + \cdots) \\ &\quad + 2(v_0 + pv_1 + p^2 v_2 + \cdots)\frac{\partial}{\partial x}(v_0 + pv_1 + p^2 v_2 + \cdots) \\ &\quad - \frac{\partial}{\partial x}((u_0 + pu_1 + p^2 u_2 + \cdots)(v_0 + pv_1 + p^2 v_2 + \cdots))\Big], \end{aligned} \right\} \tag{12.38}$$

By equating identical powers of p from both sides of Eqs. (12.37) and (12.38), one can obtain

$$p^0 : \frac{\partial u_0}{\partial t} = \frac{\partial}{\partial t} u_0(x, 0) \tag{12.39}$$

$$\frac{\partial v_0}{\partial t} = \frac{\partial}{\partial t} v_0(x, 0) \tag{12.40}$$

After integrating, these lead to

$$u_0(x, t) = u_0(x, 0) = \sin x,$$
$$v_0(x, t) = v_0(x, 0) = \sin x, \tag{12.41}$$

$$p^1 : \frac{\partial u_1}{\partial t} = \frac{\partial^2}{\partial x^2} u_0 + 2u_0 \frac{\partial u_0}{\partial x} - \left(u_0 \frac{\partial v_0}{\partial x} + v_0 \frac{\partial u_0}{\partial x} \right), \tag{12.42}$$

$$\frac{\partial v_1}{\partial t} = \frac{\partial^2}{\partial x^2} v_0 + 2v_0 \frac{\partial v_0}{\partial x} - \left(u_0 \frac{\partial v_0}{\partial x} + v_0 \frac{\partial u_0}{\partial x} \right), \tag{12.43}$$

$$p^2 : \frac{\partial u_2}{\partial t} = \frac{\partial^2}{\partial x^2} u_1 + 2 \left(u_0 \frac{\partial u_1}{\partial x} + u_1 \frac{\partial u_0}{\partial x} \right)$$
$$- \left(u_0 \frac{\partial v_1}{\partial x} + u_1 \frac{\partial v_0}{\partial x} + v_0 \frac{\partial u_1}{\partial x} + v_1 \frac{\partial u_0}{\partial x} \right), \tag{12.44}$$

$$\frac{\partial v_2}{\partial t} = \frac{\partial^2}{\partial x^2} v_1 + 2 \left(v_0 \frac{\partial v_1}{\partial x} + v_1 \frac{\partial v_0}{\partial x} \right)$$
$$- \left(u_0 \frac{\partial v_1}{\partial x} + u_1 \frac{\partial v_0}{\partial x} + v_0 \frac{\partial u_1}{\partial x} + v_1 \frac{\partial u_0}{\partial x} \right), \tag{12.45}$$

and so on.

By integrating Eqs. (12.42)–(12.45) with respect to time, we obtain

$$u_1 = \int_0^t \left(\frac{\partial^2}{\partial x^2} u_0 + 2u_0 \frac{\partial u_0}{\partial x} - \left(u_0 \frac{\partial v_0}{\partial x} + v_0 \frac{\partial u_0}{\partial x} \right) \right) dt = -t \sin x \tag{12.46}$$

$$v_1 = \int_0^t \left(\frac{\partial^2}{\partial x^2} v_0 + 2v_0 \frac{\partial v_0}{\partial x} - \left(u_0 \frac{\partial v_0}{\partial x} + v_0 \frac{\partial u_0}{\partial x} \right) \right) dt = -t \sin x \tag{12.47}$$

$$u_2 = \int_0^t \left(\frac{\partial^2}{\partial x^2} u_1 + 2 \left(u_0 \frac{\partial u_1}{\partial x} + u_1 \frac{\partial u_0}{\partial x} \right) \right.$$
$$\left. - \left(u_0 \frac{\partial v_1}{\partial x} + u_1 \frac{\partial v_0}{\partial x} + v_0 \frac{\partial u_1}{\partial x} + v_1 \frac{\partial u_0}{\partial x} \right) \right) dt = \frac{t^2}{2} \sin x \tag{12.48}$$

$$v_2 = \int_0^t \left(\frac{\partial^2}{\partial x^2} v_1 + 2 \left(v_0 \frac{\partial v_1}{\partial x} + v_1 \frac{\partial v_0}{\partial x} \right) \right.$$
$$\left. - \left(u_0 \frac{\partial v_1}{\partial x} + u_1 \frac{\partial v_0}{\partial x} + v_0 \frac{\partial u_1}{\partial x} + v_1 \frac{\partial u_0}{\partial x} \right) \right) dt = \frac{t^2}{2} \sin x. \qquad (12.49)$$

Similarly, we may obtain $u_3, u_4, \ldots$ and $v_3, v_4, \ldots$

By setting $p = 1$ in Eqs. (12.35) and (12.36), the solutions of Eqs. (12.30) and (12.31) may be found as

$$u(x, t) = \sin x - t \sin x + \frac{t^2}{2} \sin x + \cdots = \sin x \left(1 - t + \frac{t^2}{2} - \cdots \right) = \sin x e^{-t}.$$

$$v(x, t) = \sin x - t \sin x + \frac{t^2}{2} \sin x + \cdots = \sin x \left(1 - t + \frac{t^2}{2} - \cdots \right) = \sin x e^{-t}.$$

The test problems demonstrated above confirm that the HPM is an efficient method for solving linear/nonlinear ordinary, partial, and coupled differential equations. The series usually converges with an increase in the number of terms, but one may not expect the compact form solution always.

Exercise

1 Apply the HPM to find the solution differential equation $\frac{dy}{dt} + y^2 = 1$, $y(0) = 0$.

2 Solve the homogenous KdV equation $u_t - 6uu_x + u_{xxx} = 0$ subject to initial condition $u(x, 0) = 6x$ using the HPM.

References

1 He, J.H. (1999). Homotopy perturbation technique. *Computer Methods in Applied Mechanics and Engineering* 178 (3): 257–262.

2 Liu, G.L., (1997). New research directions in singular perturbation theory: artificial parameter approach and inverse-perturbation technique. Proceedings of the 7th Conference of modern Mathematics and Mechanics, Shanghai (September 1997), pp. 47–53.

3 Karunakar, P. and Chakraverty, S. (2017). Comparison of solutions of linear and non-linear shallow water wave equations using homotopy perturbation method. *International Journal of Numerical Methods for Heat and Fluid Flow* 27 (9): 2015–2029.

4 Karunakar, P. and Chakraverty, S. (2018). Solution of interval shallow water wave equations using homotopy perturbation method. *Engineering Computations* 35 (4): 1610–1624.

5 He, J.H. (2003). Homotopy perturbation method: a new nonlinear analytical technique. *Applied Mathematics and Computation* 135 (1): 73–79.
6 He, J.H. (2004). The homotopy perturbation method for nonlinear oscillators with discontinuities. *Applied Mathematics and Computation* 151 (1): 287–292.
7 Liao, S.J. (1995). An approximate solution technique not depending on small parameters: a special example. *International Journal of Non-Linear Mechanics* 30 (3): 371–380.
8 Liao, S.J. (1997). Boundary element method for general nonlinear differential operators. *Engineering Analysis with Boundary Elements* 20 (2): 91–99.
9 He, J.-H. (1999). Variational iteration method–a kind of non-linear analytical technique: some examples. *International Journal of Non-Linear Mechanics* 34 (4): 699–708.
10 Wazwaz, A.M. (2010). *Partial Differential Equations and Solitary Waves Theory*. Beijing: Springer Science & Business Media, Higher Education Press.
11 Sweilam, N.H. and Khader, M.M. (2009). Exact solutions of some coupled nonlinear partial differential equations using the homotopy perturbation method. *Computers and Mathematics with Applications* 58 (11): 2134–2141.

13

Variational Iteration Method

13.1 Introduction

The variational iteration method (VIM) is one of the well-known semi-analytical methods for solving linear and nonlinear ordinary as well as partial differential equations. He [1, 2] developed the VIM and successfully applied to ordinary and partial differential equations. Further, the method was used by many researchers for solving linear, nonlinear, homogeneous, and inhomogeneous differential equations. The main advantage of the method lies in its flexibility and ability to solve nonlinear equations easily. The method can be used in bounded and unbounded domains as well. By this method one can find the convergent successive approximations of the exact solution of the differential equations if such a solution exists. Wazwaz [3] used the VIM for solving the linear and nonlinear Volterra integral and integro-differential equations and explained clearly how to use this method for solving homogenous and inhomogeneous partial differential equations in Ref. [4].

13.2 VIM Procedure

In order to illustrate the VIM, we address the general nonlinear system [1, 2, 5] as

$$Lu + Nu = g(x, t), \tag{13.1}$$

where L is the linear operator, N is the nonlinear operator, and $g(x, t)$ is the given continuous function.

The correction functional for Eq. (13.1) may directly be written as [1, 2, 5]

$$u_{n+1}(x, t) = u_n(x, t) + \int_0^t \lambda \{Lu_n(x, \xi) + N\tilde{u}_n(x, \xi) - g(x, \xi)\} d\xi, \quad n \geq 0 \tag{13.2}$$

Advanced Numerical and Semi-Analytical Methods for Differential Equations, First Edition.
Snehashish Chakraverty, Nisha Rani Mahato, Perumandla Karunakar, and Tharasi Dilleswar Rao.
© 2019 John Wiley & Sons, Inc. Published 2019 by John Wiley & Sons, Inc.

where λ is Lagrange multiplier, which can be identified optimally via the variational theory [1, 2, 5]. Here, $\tilde{u}_n$ is a restricted variation [1, 2, 5] which means $\delta\tilde{u}_n = 0$ and u_n is the nth approximate solution. The initial approximation u_0 can be chosen freely if it satisfies the given conditions. The solution is approximated as

$$u(x, t) = \lim_{n\to\infty} u_n(x, t). \tag{13.3}$$

13.3 Numerical Examples

In this section, few test problems have been solved using the present method to make the readers familiar with the method. As such, two linear nonhomogeneous partial differential equations are handled in Examples 13.1 and 13.2, and a nonlinear partial differential equation has been solved in Example 13.3.

Example 13.1 We take a linear nonhomogeneous partial differential equation as

$$u_x - u_y = 2, \tag{13.4}$$

subject to conditions $u(x, 0) = x$, $u(0, y) = -y$.

Solution The correction functional for Eq. (13.4) may be obtained from Eq. (13.2) as (in x-direction)

$$u_{n+1}(x, y) = u_n(x, y) + \int_0^x \lambda \left(\frac{\partial}{\partial\xi} u_n(\xi, y) - \frac{\partial}{\partial y} u_n(\xi, y) - 2 \right) d\xi, \quad n \geq 0 \tag{13.5}$$

with stationary conditions [1, 2, 5]

$$1 + \lambda|_{\xi=x} = 0,$$
$$\lambda'|_{\xi=x} = 0$$

which give $\lambda = -1$.

By substituting the value of the Lagrange multiplier $\lambda = -1$ in Eq. (13.5), we get

$$u_{n+1}(x, y) = u_n(x, y) - \int_0^x \left(\frac{\partial}{\partial\xi} u_n(\xi, y) - \frac{\partial}{\partial y} u_n(\xi, y) - 2 \right) d\xi, \quad n \geq 0 \tag{13.6}$$

Next, we select $u_0(x, y) = u(0, y) = -y$ from the given conditions. By using $u_0(x, y) = -y$ in Eq. (13.6), we obtain successive approximations as below

$$u_0(x, y) = -y$$

$$u_1(x,y) = u_0(x,y) - \int_0^x \left(\frac{\partial}{\partial \xi} u_0(\xi,y) - \frac{\partial}{\partial y} u_0(\xi,y) - 2 \right) d\xi$$

$$= -y - \int_0^x \left(\frac{\partial}{\partial \xi}(-y) - \frac{\partial}{\partial y}(-y) - 2 \right) d\xi$$

$$= -y - \int_0^x (0 + 1 - 2)d\xi$$

$$= -y - \int_0^x (-1)d\xi = -y + (\xi)_0^x$$

$$= -y + (\xi)_0^x$$

$$= -y + x,$$

$$u_2(x,y) = u_1(x,y) - \int_0^x \left(\frac{\partial}{\partial \xi} u_1(\xi,y) - \frac{\partial}{\partial y} u_1(\xi,y) - 2 \right) d\xi$$

$$= -y + x - \int_0^x \left(\frac{\partial}{\partial \xi}(-y + \xi) - \frac{\partial}{\partial y}(-y + \xi) - 2 \right) d\xi$$

$$= -y + x - \int_0^x (1 + 1 - 2)d\xi$$

$$= -y + x,$$

$$u_3(x,y) = -y + x,$$

$$\vdots$$

$$u_n(x,y) = -y + x.$$

The solution of Eq. (13.4) subject to given conditions by the VIM is $u(x,y) = -y + x$, which is the exact solution.

Example 13.2   We now consider a linear nonhomogeneous partial differential equation with three independent variables

$$u_x + u_y + u_z = 3 \tag{13.7}$$

subject to the conditions: $u(0,y,z) = y + z$, $u(x,0,z) = x + z$, and $u(x,y,0) = x + y$.

Solution   Here, the correction functional for Eq. (13.7) may be written as (in x-direction)

$$u_{n+1}(x,y,z) = u_n(x,y,z) + \int_0^x \lambda \left(\frac{\partial}{\partial \xi} u_n(\xi,y,z) + \frac{\partial}{\partial y} u_n(\xi,y,z) \right.$$

$$\left. + \frac{\partial}{\partial z} u_n(\xi,y,z) - 3 \right) d\xi, \quad n \geq 0 \ldots \tag{13.8}$$

with stationary conditions [1, 2, 5]

$$1 + \lambda|_{\xi=x} = 0,$$

$$\lambda'|_{\xi=x} = 0$$

which yield $\lambda = -1$.

By substituting the above Lagrange multiplier $\lambda = -1$ in Eq. (13.8), we have

$$u_{n+1}(x, y, z) = u_n(x, y, z) - \int_0^x \left(\frac{\partial}{\partial \xi} u_n(\xi, y, z) + \frac{\partial}{\partial y} u_n(\xi, y, z) \right.$$
$$\left. + \frac{\partial}{\partial z} u_n(\xi, y, z) - 3 \right) d\xi, \quad n \geq 0 \dots. \tag{13.9}$$

Now, we select $u_0(x, y, z) = u(0, y, z) = y + z$ from the given conditions. Using this in Eq. (13.9), we obtain successive approximations as

$$u_0(x, y, z) = y + z,$$

$$u_1(x, y, z) = u_0(x, y, z) - \int_0^x \left(\frac{\partial}{\partial \xi} u_0(\xi, y, z) + \frac{\partial}{\partial y} u_0(\xi, y, z) + \frac{\partial}{\partial z} u_0(\xi, y, z) - 3 \right) d\xi$$

$$= y + z - \int_0^x \left(\frac{\partial}{\partial \xi} (y + z) + \frac{\partial}{\partial y} (y + z) + \frac{\partial}{\partial z} (y + z) - 3 \right) d\xi$$

$$= y + z - \int_0^x (0 + 1 + 1 - 3) d\xi,$$

$$= y + z - \int_0^x (-1) d\xi = y + z + (\xi)_0^x$$

$$= y + z + (\xi)_0^x$$

$$= x + y + z,$$

$$u_2(x, y, z) = u_1(x, y, z) - \int_0^x \left(\frac{\partial}{\partial \xi} u_1(\xi, y, z) + \frac{\partial}{\partial y} u_1(\xi, y, z) + \frac{\partial}{\partial z} u_1(\xi, y, z) - 3 \right) d\xi$$

$$= x + y + z - \int_0^x \left(\frac{\partial}{\partial \xi} (\xi + y + z) + \frac{\partial}{\partial y} (\xi + y + z) + \frac{\partial}{\partial z} (\xi + y + z) - 3 \right) d\xi$$

$$= x + y + z - \int_0^x (1 + 1 + 1 - 3) d\xi$$

$$= x + y + z,$$

$$u_3(x, y, z) = x + y + z,$$

$$\vdots$$

$$u_n(x, y, z) = x + y + z.$$

Thus, the exact solution of Eq. (13.7) by the VIM is $u(x, y, z) = x + y + z$.

Example 13.3   Finally, we consider a nonlinear nonhomogeneous partial differential equation

$$u_t + u u_x = x \tag{13.10}$$

subject to the initial condition $u(x, 0) = 2$.

Solution   The correction functional for Eq. (13.10) is (in t-direction)

$$u_{n+1}(x, t) = u_n(x, t) + \int_0^t \lambda \left(\frac{\partial}{\partial \xi} u_n(x, \xi) + u_n(x, \xi) \frac{\partial}{\partial x} u_n(x, \xi) - x \right) d\xi, \quad n \geq 0 \ldots$$

$$(13.11)$$

Here, the stationary conditions [1, 2, 5] are

$$1 + \lambda|_{\xi=x} = 0,$$
$$\lambda'|_{\xi=x} = 0$$

from which we get $\lambda = -1$.

Substituting Lagrange multiplier $\lambda = -1$ in Eq. (13.11) results in

$$u_{n+1}(x, t) = u_n(x, t) - \int_0^t \left(\frac{\partial}{\partial \xi} u_n(x, \xi) + u_n(x, \xi) \frac{\partial}{\partial x} u_n(x, \xi) - x \right) d\xi, \quad n \geq 0$$

$$(13.12)$$

Now, we select $u_0(x, t) = u(x, 0) = 2$ from the given conditions. Using this in Eq. (13.12), we obtain successive approximations as

$$u_0(x, t) = 2$$

$$u_1(x, t) = u_n(x, t) - \int_0^t \left(\frac{\partial}{\partial \xi} u_0(x, \xi) + u_0(x, \xi) \frac{\partial}{\partial x} u_0(x, \xi) - x \right) d\xi$$

$$= 2 - \int_0^t \left(\frac{\partial}{\partial \xi}(2) + 2 \frac{\partial}{\partial x}(2) - x \right) d\xi$$

$$= 2 - \int_0^t (-x) d\xi$$

$$= 2 + xt$$

$$u_2(x, t) = u_1(x, t) - \int_0^t \left(\frac{\partial}{\partial \xi} u_1(x, \xi) + u_1(x, \xi) \frac{\partial}{\partial x} u_1(x, \xi) - x \right) d\xi$$

$$= 2 + xt - \int_0^t \left(\frac{\partial}{\partial \xi}(2 + x\xi) + (2 + x\xi) \frac{\partial}{\partial x}(2 + x\xi) - x \right) d\xi$$

$$= 2 + xt - \int_0^t (x + (2 + x\xi)\xi - x) d\xi$$

$$= 2 + xt - \int_0^t (2\xi + x\xi^2) d\xi$$

$$= 2 + xt - t^2 - \frac{1}{3} xt^3$$

$$u_3(x, t) = u_2(x, t) - \int_0^t \left(\frac{\partial}{\partial \xi} u_2(x, \xi) + u_2(x, \xi) \frac{\partial}{\partial x} u_2(x, \xi) - x \right) d\xi$$

$$= 2 + xt - t^2 - \frac{1}{3} xt^3$$

$$-\int_0^t \left(\frac{\partial}{\partial\xi}\left(2+x\xi-\xi^2-\frac{1}{3}x\xi^3\right)+\left(2+x\xi-\xi^2-\frac{1}{3}x\xi^3\right)\right.$$

$$\times \frac{\partial}{\partial x}\left(2+x\xi-\xi^2-\frac{1}{3}x\xi^3\right)-x\right)d\xi$$

$$=2+xt-t^2-\frac{1}{3}xt^3-\int_0^t\left(x-2\xi-x\xi^2+\left(2+x\xi-\xi^2-\frac{1}{3}x\xi^3\right)\right.$$

$$\times\left(\xi-\frac{1}{3}\xi^3\right)-x\right)d\xi$$

$$=2+xt-t^2-\frac{1}{3}xt^3-\int_0^t\left(-\frac{5}{3}\xi^3-\frac{2}{3}x\xi^4+\frac{1}{3}\xi^5-\frac{1}{3}x\xi^6\right)d\xi$$

$$=2+xt-t^2-\frac{1}{3}xt^3-\left(-\frac{5}{12}\xi^4-\frac{2}{15}x\xi^5+\frac{1}{18}\xi^6+\frac{1}{21}x\xi^7\right)_0^t$$

$$=2+xt-t^2-\frac{1}{3}xt^3+\frac{5}{12}t^4+\frac{2}{15}xt^5-\frac{1}{18}t^6-\frac{1}{21}xt^7$$

and so on.

In general, we have

$$u_n(x,t)=2+xt-t^2-\frac{1}{3}xt^3+\frac{5}{12}t^4+\frac{2}{15}xt^5-\frac{61}{360}t^6-\frac{17}{315}xt^7\cdots$$

$$=2\left(1-\frac{1}{2}t^2+\frac{5}{24}t^4-\frac{61}{720}t^6\cdots\right)+x\left(t-\frac{1}{3}t^3+\frac{2}{15}t^5-\frac{17}{315}xt^7\cdots\right)$$

$$=2\,\mathrm{sech}\,x+x\tanh x.$$

As such, the exact solution of the nonlinear differential equation is $u(x,t)=2\,\mathrm{sech}\,x+x\tanh x$.

Exercise

1 Apply the VIM to find the solution differential equation $u_t-uu_x=0$, subject to initial condition $u(x,0)=\frac{x}{10}$.

2 Solve $u_t+uu_x=2x, x\in\mathfrak{R}, t\geq 0$ subject to initial condition $u(x,0)=x$ using the VIM.

3 Solve the homogenous Burger's equation $u_t+uu_x-u_{xx}=0$ subject to initial condition $u(x,0)=x$ using the VIM.

References

1 He, J.H. (1998). Approximate analytical solution for seepage flow with fractional derivatives in porous media. *Computer Methods in Applied Mechanics and Engineering* 167 (1–2): 57–68.

2 He, J.H. and Wu, X.H. (2007). Variational iteration method: new development and applications. *Computers and Mathematics with Applications* 54 (7/8): 881–894.

3 Wazwaz, A.M. (2010). The variational iteration method for solving linear and nonlinear Volterra integral and integro-differential equations. *International Journal of Computer Mathematics* 87 (5): 1131–1141.

4 Wazwaz, A.M. (2010). *Partial Differential Equations and Solitary Waves Theory*. Beijing: Springer Science & Business Media, Higher Education Press.

5 He, J.H. (2000). Variational iteration method for autonomous ordinary differential systems. *Applied Mathematics and Computation* 114 (2): 115–123.

14

Homotopy Analysis Method

14.1 Introduction

Homotopy analysis method (HAM) [1–7] is one of the well-known semi-analytical methods for solving various types of linear and nonlinear differential equations (ordinary as well as partial). This method is based on coupling of the traditional perturbation method and homotopy in topology. By this method one may get exact solution or a power series solution which converges in general to exact solution. The HAM consists of parameter $\hbar \neq 0$ called as convergence control parameter, which controls the convergent region and rate of convergence of the series solution. This method was first proposed by Liao [4]. The same was successfully employed to solve many types of problems in science and engineering [1–8] and the references mentioned therein.

14.2 HAM Procedure

To illustrate the idea of HAM, we consider the following differential equation [1–7] in general

$$N(u(x,t)) = 0, \quad (x,t) \in \Omega \tag{14.1}$$

where N is the operator (linear or nonlinear) and u is the unknown function in the domain Ω.

The method begins by defining homotopy operator H as below [7],

$$H(\phi, p) \equiv (1-p)[L(\phi(x;p)) - u_0(x)] - p\hbar N(\phi(x;p)), \tag{14.2}$$

where $p \in [0, 1]$ is an embedding parameter and $\hbar \neq 0$ is the convergence control parameter [3, 7], u_0 is an initial approximation of the solution of Eq. (14.1), ϕ

Advanced Numerical and Semi-Analytical Methods for Differential Equations, First Edition.
Snehashish Chakraverty, Nisha Rani Mahato, Perumandla Karunakar, and Tharasi Dilleswar Rao.
© 2019 John Wiley & Sons, Inc. Published 2019 by John Wiley & Sons, Inc.

is an unknown function, and L is the auxiliary linear operator satisfying the property $L(0) = 0$.

By considering $H(\phi, p) = 0$, we obtain

$$(1 - p)[L(\phi(x, t; p)) - u_0(x, t)] = p\hbar N(\phi(x, t; p)) \tag{14.3}$$

which is called the *zero-order deformation equation*.

From Eq. (14.3), it is clear that for $p = 0$ we get $L(\phi(x, t; 0)) - u_0(x, t) = 0$ that gives $\phi(x; 0) = u_0(x, t)$. On the other hand, for $p = 1$, Eq. (14.3) reduces to $N(\phi(x, t; p)) = 0$ which gives $\phi(x, t; 1) = u(x)$. So, by changing p from 0 to 1 the solution changes from u_0 to u.

Using Maclaurin series, the function $\phi(x, t; p)$ with parameter p may be written as [7]

$$\phi(x, t; p) = \phi(x, t; 0) + \sum_{m=1}^{\infty} \frac{1}{m!} \left. \frac{\partial^m \phi(x, t; p)}{\partial p^m} \right|_{p=0} p^m. \tag{14.4}$$

Denoting

$$u_m(x, t) = \frac{1}{m!} \left. \frac{\partial^m \phi(x, t; p)}{\partial p^m} \right|_{p=0}, \quad m = 1, 2, 3, \ldots, \tag{14.5}$$

Eq. (14.4) turns into

$$\phi(x, t; p) = u_0(x, t) + \sum_{m=1}^{\infty} u_m(x, t) p^m. \tag{14.6}$$

If the series (14.6) converges for $p = 1$ [7], then we obtain the solution of Eq. (14.1) as

$$u(x, t) = \sum_{m=0}^{\infty} u_m(x, t). \tag{14.7}$$

In order to determine the function u_m, we differentiate Eq. (14.3), m times with respect to p. Next we divide the result by $m!$ and substitute $p = 0$ [7]. In this way we may find the mth-order deformation equation for $m > 0$ as below [7]

$$L(u_m(x, t) - \chi_m u_{m-1}(x, t)) = \hbar H(x, t) R_m(u_{m-1}(x, t)) \tag{14.8}$$

where $H(x, t)$ is the auxiliary function, $\chi_m = \begin{cases} 0, & \text{if } m \leq 1 \\ 1, & \text{if } m > 1 \end{cases}$ and

$$R_m(u_{m-1}(x, t)) = \frac{1}{(m-1)!} \left(\frac{\partial^{m-1}}{\partial p^{m-1}} N\left(\sum_{i=1}^{\infty} u_i(x) p^i \right) \right). \tag{14.9}$$

14.3 Numerical Examples

Here, we apply the present method to solve a linear partial differential equation in Example 14.1 and a nonlinear partial differential equation in Example 14.2.

Example 14.1   We consider a homogeneous linear partial differential equation from

$$2u_t + u_x - u_{xxt} = 0 \tag{14.10}$$

subject to the initial condition $u(x, 0) = e^{-x}$.

Solution   Here, the auxiliary linear operator is $\dfrac{\partial}{\partial t}$.

Using the above-discussed HAM, the zeroth-order deformation equation is (from Eq. (14.3))

$$(1 - p)\left(\frac{\partial}{\partial t}(\phi(x, t; p)) - u(x, 0)\right)$$

$$= p\hbar\left(2\frac{\partial}{\partial t}(\phi(x, t; p)) + \frac{\partial}{\partial t}(\phi(x, t; p)) - \frac{\partial^3}{\partial x^2 \partial t}(\phi(x, t; p))\right) \tag{14.11}$$

subject to the initial condition

$$\phi(x, 0, p) = e^{-x}.$$

where $p \in [0, 1]$. Obviously, if we take $p = 0$ we get $\phi(x, t; 0) = e^{-x}$ and when $p = 1$ the equation is equivalent to Eq. (14.10). Thus, it holds

$$\phi(x, t; 1) = u(x, t),$$

where

$$\phi = \phi(x, t; p) = u_0(x, t) + \sum_{m=1}^{\infty} u_m(x, t)p^m. \tag{14.12}$$

The mth-order deformation equation is [5, 7, 8]

$$L(u_m(x, t) - \chi_m u_{m-1}(x, t)) = \hbar H(x, t) R_m(u_{m-1}(x, t)) \tag{14.13}$$

subject to

$$u_m(x, 0) = 0$$

where

$$\chi_m = \begin{cases} 0, & \text{if } m \le 1 \\ 1, & \text{if } m > 1 \end{cases}$$

and

$$R_m(u_{m-1}(x, t)) = \left(2\frac{\partial}{\partial t}u_{m-1} + \frac{\partial}{\partial x}u_{m-1} - \frac{\partial^3}{\partial x^2 \partial t}u_{m-1}\right). \tag{14.14}$$

Let $H(x, t) = 1$, then Eq. (14.13) may be rewritten as

$$L(u_m(x, t) - \chi_m u_{m-1}(x, t)) = \hbar R_m(u_{m-1}(x, t))$$

$$u_m(x, t) - \chi_m u_{m-1}(x, t) = \hbar L^{-1}(R_m(u_{m-1}(x, t)))$$

where L^{-1} is the inverse linear differential operator.

$$u_m(x, t) = \chi_m u_{m-1}(x, t) + \hbar \int_0^t R_m(u_{m-1}(x, t)) d\tau$$

Let $u_0(x, t) = u(x, 0) = e^{-x}$

$$u_1(x, t) = \chi_1 u_0(x, t) + \hbar \int_0^t R_0(u_0(x, t) \, d\tau$$

$$= 0 + \hbar \int_0^t \left(2\frac{\partial}{\partial \tau}(e^{-x}) + \frac{\partial}{\partial x}(e^{-x}) - \frac{\partial^2}{\partial x^2 \partial \tau}(e^{-x}) \right) d\tau$$

$$= 0 + \hbar \int_0^t (0 - e^{-x} - 0) \, d\tau$$

$$= -\hbar e^{-x} t$$

$$u_2(x, t) = \chi_2 u_1(x, t) + \hbar \int_0^t R_1(u_1(x, t)) \, d\tau$$

$$= -\hbar e^{-x} t + \hbar \int_0^t \left(2\frac{\partial}{\partial \tau}(-\hbar e^{-x} \tau) + \frac{\partial}{\partial x}(-\hbar e^{-x} \tau) \right.$$

$$\left. - \frac{\partial^2}{\partial x^2 \partial \tau}(-\hbar e^{-x} \tau) \right) d\tau$$

$$= -\hbar e^{-x} t + \hbar \int_0^t (-2\hbar e^{-x} + \hbar e^{-x} \tau + \hbar e^{-x}) \, d\tau$$

$$= -\hbar e^{-x} t + \hbar \left(-2\hbar e^{-x} t + \hbar e^{-x}\frac{t^2}{2} + \hbar e^{-x} t \right)$$

$$= \frac{\hbar e^{-x} t}{2}(\hbar t - 2\hbar - 2)$$

$$u_3(x, t) = \chi_3 u_2(x, t) + \hbar \int_0^t R_2(u_2(x, t)) \, d\tau$$

$$= \frac{\hbar e^{-x} t}{2}(\hbar t - 2\hbar - 2)$$

$$+ \hbar \int_0^t \left(2\frac{\partial}{\partial \tau}(u_2(x, \tau)) + \frac{\partial}{\partial x}(u_2(x, \tau)) - \frac{\partial^2}{\partial x^2 \partial \tau}(u_2(x, \tau)) \right) d\tau$$

$$= \frac{\hbar e^{-x} t}{6}(-\hbar^2 t^2 + 6\hbar t + 6\hbar^2 t - 6\hbar^2 - 12\hbar + 6).$$

$$\vdots$$

The final solution of Eq. (14.10) is

$$u(x, t) = u_0(x, t) + u_1(x, t) + u_2(x, t) + u_3(x, t) \cdots$$

$$= e^{-x} - \hbar e^{-x}t + \frac{\hbar e^{-x}t}{2}(\hbar t - 2\hbar - 2)$$

$$+ \frac{\hbar e^{-x}t}{6}(-\hbar^2 t^2 + 6\hbar t + 6\hbar^2 t - 6\hbar^2 - 12\hbar + 6) + \cdots$$

$$= e^{-x}\left(1 - \hbar t + \frac{\hbar t}{2}(\hbar t - 2\hbar - 2)\right.$$

$$\left. + \frac{\hbar t}{6}(-\hbar^2 t^2 + 6\hbar t + 6\hbar^2 t - 6\hbar^2 - 12\hbar + 6) + \cdots\right)$$

The approximate solution plots of Eq. (14.10) of order 6 ($u(x, t) = u_0$ $(x, t) + u_1(x, t) + u_2(x, t) + u_3(x, t) + u_4(x, t) + u_5(x, t) + u_6(x, t))$ for $\hbar = -0.5, -0.8, -1.0, -1.2$ are depicted in Figure 14.1 and also one may see the comparison of present results with exact solution at $t = 1$. Similarly Figure 14.2 represents the same type of results for $\hbar = -0.5, -0.7, -0.8, -1.0$ at $t = 2$.

Here, the exact solution of Eq. (14.10) is $u(x, t) = e^{-x+t}$.

From the results presented in Figure 14.1, one may see that for $\hbar = -1.0$ the HAM solution is exactly the same as exact solution. The solution for $\hbar = -0.80$ is also converging to exact solution. Figure 14.2 also confirms that for $t = 2$, the HAM solution is converging to exact solution for $\hbar = -1.2$. From the above discussion it may be concluded that proper values of control parameter $\hbar$ should be chosen for getting better solution with less error. It may be noted that the HAM not only produces approximate convergent series solution but it can also give exact solution depending on the considered problem with proper $\hbar$. This may be observed in the next example.

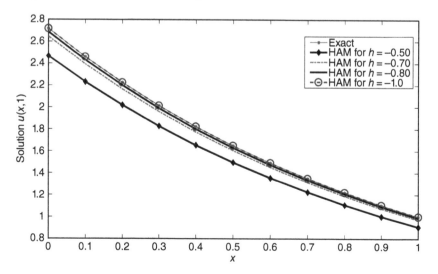

Figure 14.1 Solution of Eq. (14.10) for different values of $\hbar$ at $t = 1$.

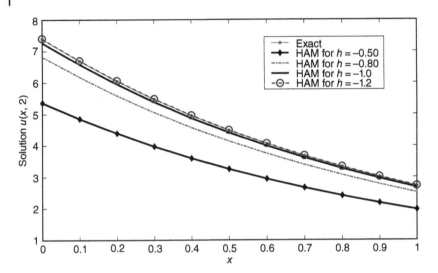

Figure 14.2 Solution of Eq. (14.10) for different values of $\hbar$ at $t = 2$.

Example 14.2 Now, we take a nonlinear and nonhomogeneous advection equation [9] as

$$u_t + uu_x = x \tag{14.15}$$

subject to the initial condition $u(x, 0) = 2$.

The zeroth order deformation equation in this case is

$$(1 - p)\left(\frac{\partial}{\partial t}(\phi(x, t; p)) - u(x, 0)\right)$$
$$= p\hbar\left(\frac{\partial}{\partial t}(\phi(x, t; p)) + \phi(x, t; p)\frac{\partial}{\partial x}(\phi(x, t; p)) - x\right) \tag{14.16}$$

subject to

$$\phi = \phi(x, 0; p) = 2,$$

where $p \in [0, 1]$. Obviously, if we take $p = 0$ we get $(x, t; 0) = 2$ and when $p = 1$ the equation is equivalent to Eq. (14.15). Thus, it holds $\phi(x, t; 1) = u(x, t)$.

The Taylor series expansion of $\phi(x, t; p)$ is

$$\phi = \phi(x, t; p) = u_0(x, t) + \sum_{m=1}^{\infty} u_m(x, t)p^m \tag{14.17}$$

By substituting Eq. (14.17) into Eq. (14.16) and equating the coefficients of like powers of p, the mth-order deformation equation may be obtained as below [8]

$$L(u_m(x, t) - \chi_m u_{m-1}(x, t)) = \hbar H(x, t)R_m(u_{m-1}(x, t)) \tag{14.18}$$

Subject to $u_m(x, 0) = 0$

For $H(x, t) = 1$, the solution of the mth-order deformation equation (14.18) reads as

$$u_m(x, t) = \chi_m u_{m-1}(x, t) + \hbar \int_0^t R_m(u_{m-1}(x, t)) \, d\tau \tag{14.19}$$

where

$$\chi_m = \begin{cases} 0, & \text{if } m \leq 1 \\ 1, & \text{if } m > 1 \end{cases}$$

and

$$R_m(u_{m-1}(x, t)) = \left(\frac{\partial}{\partial t} u_{m-1} + u_{m-1} \frac{\partial}{\partial x} u_{m-1} - x \right) \tag{14.20}$$

Let $u_0(x, t) = u(x, 0) = 2$

$$u_1(x, t) = 0 + \hbar \int_0^t R_0(u_0(x, t)) \, d\tau$$

$$= \hbar \int_0^t (-x) \, d\tau$$

$$= \hbar(-xt)$$

$$= -\hbar xt$$

$$u_2(x, t) = u_1(x, t) + \hbar \int_0^t R_1(u_1(x, t)) \, d\tau$$

$$= u_1(x, t) + \hbar \int_0^t \left(\frac{\partial}{\partial \tau}(u_1) + u_1 \frac{\partial}{\partial x}(u_1) - x \right) d\tau$$

$$= -\hbar xt + \hbar \int_0^t (-\hbar x + \hbar^2 x\tau^2 - x) \, d\tau$$

$$= -\hbar xt + \hbar \left(-\hbar xt + \hbar^2 x \frac{t3}{3} - xt \right)$$

So, the solution of Eq. (14.15) up to three terms is

$$u(x, t) = u_0(x, t) + u_1(x, t) + u_2(x, t) + \cdots$$

$$= 2 - \hbar xt - \hbar xt + \hbar \left(-\hbar xt + \hbar^2 x \frac{t3}{3} - xt \right) + \cdots$$

One may see that for $\hbar = -1$, $u(x, t) = 2 \left(1 - \frac{1}{2}t^2 + \frac{5}{24}t^4 - \cdots \right) + x \left(t - \frac{1}{3}t^3 + \frac{2}{15}t^5 - \cdots \right)$

$$u(x, t) = 2 \operatorname{sech} t + x \tanh t$$

which is same as the exact solution [9].

Exercise

1 Apply the HAM to find the solution of the partial differential equation $u_t + \frac{1}{4}u_x^2 = x^2$, subject to the initial condition $u(x, 0) = 0$.

2 Find the solution of $u_t + uu_t = x^2$ subject to $u(x, 0) = \frac{x}{10}$.

References

1 Wazwaz, A.M. (2008). A study on linear and nonlinear Schrodinger equations by the variational iteration method. *Chaos, Solitons and Fractals* 37 (4): 1136–1142.

2 Song, L. and Zhang, H. (2007). Application of homotopy analysis method to fractional KdV–Burgers–Kuramoto equation. *Physics Letters A* 367 (1): 88–94.

3 Liao, S. (2004). On the homotopy analysis method for nonlinear problems. *Applied Mathematics and Computation* 147 (2): 499–513.

4 Liao, S. (2012). *Homotopy Analysis Method in Nonlinear Differential Equations*, 153–165. Beijing: Higher Education Press.

5 Liang, S. and Jeffrey, D.J. (2009). Comparison of homotopy analysis method and homotopy perturbation method through an evolution equation. *Communications in Nonlinear Science and Numerical Simulation* 14 (12): 4057–4064.

6 Abbasbandy, S. (2006). The application of homotopy analysis method to nonlinear equations arising in heat transfer. *Physics Letters A* 360 (1): 109–113.

7 Hetmaniok, E., Słota, D., Trawiński, T., and Wituła, R. (2014). Usage of the homotopy analysis method for solving the nonlinear and linear integral equations of the second kind. *Numerical Algorithms* 67 (1): 163–185.

8 Liao, S.J., 1992. The proposed homotopy analysis technique for the solution of nonlinear problems. Doctoral dissertation, PhD thesis, Shanghai Jiao Tong University.

9 Wazwaz, A.M. (2010). *Partial Differential Equations and Solitary Waves Theory*. Beijing: Springer Science & Business Media, Higher Education Press.

15

Differential Quadrature Method

15.1 Introduction

We have already discussed Akbari–Ganji's method (AGM), exp-function method, Adomian decomposition method (ADM), homotopy perturbation method (HPM), variational iteration method (VIM), and homotopy analysis method (HAM) in Chapters 9, 10, 11, 12, 13, and 14, respectively. This chapter presents another effective numerical method that approximates the solution of the PDEs by functional values at certain discretized points. This method can be applied with considerably less number of grid points, whereas the methods like finite difference method (FDM), finite element method (FEM), and finite volume method (FVM) as given in Chapters 5, 6, and 7, respectively, may need more number of grid points to obtain the solution. However, FDM, FEM, and FVM are versatile methods which may certainly be used to handle regular as well as irregular domains.

Approximating partial derivatives by means of weighted sum of function values is known as Differential Quadrature (DQ). The differential quadrature method (DQM) was proposed by Bellman et al. [1–4]. Determination of weighted coefficients plays an important and crucial role in DQM. As such, an effective procedure for finding weighted coefficients suggested by Bellman is to use a simple algebraic formulation of weighted coefficient with the help of coordinates of grid points. Here, the coordinate of the grid points are the roots of the base functions like Legendre polynomials, Chebyshev polynomials, etc. Interested readers may see Refs. [5–10] and references therein for more details and application of DQM.

15.2 DQM Procedure

In this section, DQM has been illustrated [1, 2] by considering a second-order nonlinear partial differential equation (PDE) of the form

$$u_t(x, t) = f(x, t, u(x, t), u_x(x, t), u_{xx}(x, t)), \tag{15.1}$$

Advanced Numerical and Semi-Analytical Methods for Differential Equations, First Edition.
Snehashish Chakraverty, Nisha Rani Mahato, Perumandla Karunakar, and Tharasi Dilleswar Rao.
© 2019 John Wiley & Sons, Inc. Published 2019 by John Wiley & Sons, Inc.

subject to initial condition

$$u(x, 0) = g(x). \tag{15.2}$$

The DQ of the first- and second-order spatial derivatives at the grid points x_i are given by

$$u_x(x_i, t) = \sum_{j=1}^{N} a_{ij} u(x_j, t), \quad i = 1, 2, 3, \ldots, N, \tag{15.3}$$

$$u_{xx}(x_i, t) = \sum_{j=1}^{N} b_{ij} u(x_j, t), \quad i = 1, 2, 3, \ldots, N, \tag{15.4}$$

By plugging Eqs. (15.3) and (15.4) in Eq. (15.1), we obtain a system of N ordinary differential equations (ODEs) as follows:

$$u_t(x_i, t) = f\left(x_i, t, u(x_i, t), \sum_{j=1}^{N} a_{ij} u(x_j, t), \sum_{j=1}^{N} b_{ij} u(x_j, t) \right), \tag{15.5}$$

subject to initial conditions

$$u(x_i, 0) = g(x_i). \tag{15.6}$$

The weighted coefficients a_{ij} and b_{ij} are to be found using N grid points x_i and these grid points depend on the considered polynomials like Legendre, Chebyshev, etc. In this chapter, shifted Legendre polynomials $P_N^*(x_i)$ in [0, 1] have been used to find weighted coefficients.

The shifted Legendre polynomials [11, 12] of order N are given as

$$P_N^*(x_i) = \sum_{k=0}^{N} (-1)^{N+k} \frac{(N+k)!}{(N-k)!(k!)^2} x_i^k$$

First, five shifted Legendre polynomials are

$$P_0^*(x_i) = 1$$
$$P_1^*(x_i) = 2x_i - 1$$
$$P_2^*(x_i) = 6x_i^2 - 6x_i + 1$$
$$P_3^*(x_i) = 20x_i^3 - 30x_i^2 + 12x_i - 1$$
$$P_4^*(x_i) = 70x_i^4 - 140x_i^3 + 90x_i^2 - 20x_i + 1$$

The weighted coefficients in terms of shifted Legendre polynomial $P_N^*(x_i)$ [2] are taken as follows:

$$a_{ij} = \frac{P_N'^*(x_i)}{(x_i - x_j) P_N'^*(x_j)}, \quad i \neq j \tag{15.7}$$

and

$$a_{ii} = \frac{1 - 2x_i}{2x_i(x_i - 1)}, \quad i = j \tag{15.8}$$

Once all a_{ii} are obtained, it is then easy to find b_{ii} following the below procedure.

Let us denote Eq. (15.3) as

$$\frac{\partial u}{\partial x} = Au. \tag{15.9}$$

where $A = [a_{ij}]_{N \times N}$.

The second-order derivative can be approximated as

$$\frac{\partial^2 u}{\partial x^2} = \frac{\partial}{\partial x}\left(\frac{\partial u}{\partial x}\right) = \frac{\partial}{\partial x}(Au) = A\frac{\partial u}{\partial x} = A(Au) = A^2 u, \tag{15.10}$$

where $[b_{ij}] = A^2$.

Hence, weighted coefficients b_{ij} can be obtained by squaring A as

$$[b_{ij}] = [a_{ij}]_{N \times N} \times [a_{ij}]_{N \times N}.$$

Substituting the obtained weighted coefficients a_{ij} and b_{ij} in Eq. (15.5) and then using a suitable method to solve the system of ODEs, the solution of the model problem may be found at selected grid points.

Next, we solve two test problems to demonstrate the DQM.

15.3 Numerical Examples

In this section, two nonlinear PDEs are solved by considering different N for better understanding of the DQM.

Example 15.1 Let us consider the following PDE:

$$u_t(x, t) = x^2 + \frac{1}{4}u_x^2(x, t), \tag{15.11}$$

subject to the initial condition $u(x, 0) = 0$.

Solution First, we need to write derivative term as approximating sum except the derivatives with respect to t.

Hence, replacing $u_x(x, t)$ by $\sum_{j=1}^{N} a_{ij}u(x_j, t)$ in Eq. (15.11), we get a set of ODEs as

$$u_t(x_i, t) = x_i^2 + \frac{1}{4}\left[\sum_{j=1}^{N} a_{ij}u(x_j, t)\right]^2, \tag{15.12}$$

subject to the initial conditions

$$u(x_i, 0) = 0, \quad i = 1, 2, \ldots, N \tag{15.13}$$

Here, we demonstrate the procedure for $N = 3$. Then, Eq. (15.12) may be written as

$$u_t(x_i, t) = x_i^2 + \frac{1}{4}[a_{i1}u(x_1, t) + a_{i2}u(x_2, t) + a_{i3}u(x_3, t)]^2, \quad i = 1, 2, 3 \tag{15.14}$$

subject to the conditions $u(x_i, 0) = 0, \quad i = 1, 2, 3.$
The matrix form of Eq. (15.14) may be written as

$$\begin{bmatrix} u_t(x_1, t) \\ u_t(x_2, t) \\ u_t(x_3, t) \end{bmatrix} = \frac{1}{4} \left\{ \begin{bmatrix} a_{11} & a_{12} & a_{13} \\ a_{21} & a_{22} & a_{23} \\ a_{31} & a_{32} & a_{33} \end{bmatrix} \begin{bmatrix} u(x_1, t) \\ u(x_2, t) \\ u(x_3, t) \end{bmatrix} \right\}^2 + \begin{bmatrix} x_1^2 \\ x_2^2 \\ x_3^2 \end{bmatrix} \tag{15.15}$$

Now we find the weighted coefficient a_{ij} using shifted Legendre polynomial $P_3^*(x_i)$.
The grid points $x_i, i = 1, 2, 3$ are the roots of $P_3^*(x_i) = 20x_i^3 - 30x_i^2 + 12x_i - 1 = 0$ and they may be found as

$$x_1 = 0.112\,701\,665\,4$$
$$x_2 = 0.5$$

and

$$x_3 = 0.887\,298\,334\,6$$

Here, the weighted coefficients of the derivative that are found using Eqs. (15.7) and (15.8) are

$$[a_{ij}] = \begin{bmatrix} -3.872\,983\,35 & -2.581\,988\,90 & -1.290\,994\,45 \\ -2.581\,988\,90 & 0 & -2.581\,988\,90 \\ 1.290\,994\,45 & -2.581\,988\,90 & -3.872\,983\,35 \end{bmatrix}$$

and

$$[x_i^2] = \begin{bmatrix} 0.012\,701\,67 \\ 0.250\,000\,00 \\ 0.787\,298\,33 \end{bmatrix}, \quad i = 1, 2, 3$$

The system of nonlinear ODEs (15.12) along with initial conditions has been solved using the fourth-order Runge–Kutta method and the results are reported in Table 15.1.
Here, the exact solution [13] of Eq. (15.11) is

$$u(x, t) = x^2 \tan(t). \tag{15.16}$$

Table 15.1 Comparison of solutions of PDE (15.11) by DQM for $N = 3$ with the exact solution.

t	x	DQM solution	Exact solution
0.01	x_1	0.00012726	0.00012702
	x_2	0.00250033	0.00250008
	x_3	0.00787413	0.00787325
0.1	x_1	0.00151955	0.00127442
	x_2	0.02533779	0.02508367
	x_3	0.07989748	0.07899332

From the results presented in Table 15.1 it may be confirmed that solutions at particular nodes obtained by the DQM are in good agreement with that of the exact solution. It may be worth mentioning that we have considered here a small N value ($N = 3$) so that readers can understand the method well. However, better results may be obtained by taking a large value of N.

This effect of N is shown in Example 15.2 by considering two different N values for the same problem.

Example 15.2 Let us consider now a nonhomogeneous advection equation from Ref. [10] as

$$u_t + uu_x = x \tag{15.17}$$

subject to the initial condition: $u(x, 0) = 2$.

Solution The exact solution of Eq. (15.17) is $u = 2 \operatorname{sech} t + x \tanh t$ [10].

Now we solve PDE (15.17), with the given initial condition using the DQM for $N = 4$ and $N = 7$.

By replacing $u_x(x, t)$ by $\sum_{j=1}^{N} a_{ij} u(x_j, t)$ in Eq. (15.17), we get a set of nonlinear system of ODEs as

$$u_t(x_i, t) = -u(x_i, t) \sum_{j=1}^{N} a_{ij} u(x_j, t) + x_i \tag{15.18}$$

The grid points of $P_N^*(x)$ for $N = 4$ may be obtained by following the previously mentioned procedure as

$$x_1 = 0.069\,431\,844\,202\,974;$$
$$x_2 = 0.330\,094\,782\,075\,720;$$
$$x_3 = 0.669\,990\,521\,792\,431;$$
$$x_4 = 0.930\,568\,155\,790\,230;$$

The weighted coefficients using Eqs. (15.7) and (15.8) are found as

$$[a_{ij}] = \begin{bmatrix} -6.664\,000\,47 & -3.836\,372\,01 & -1.665\,116\,23 & -1.161\,256\,34 \\ 3.836\,372\,02 & -0.768\,342\,22 & -2.942\,078\,65 & -1.665\,352\,78 \\ 1.665\,116\,23 & 2.942\,078\,65 & 0.768\,828\,78 & -3.837\,627\,91 \\ 1.161\,256\,34 & 1.665\,352\,78 & 3.837\,627\,91 & 6.664\,000\,47 \end{bmatrix}$$

By plugging weighted coefficients a_{ij} and nodes x_i in Eq. (15.18) and solving nonlinear system of differential equations using the Runge–Kutta method of order four, we may get the solution of the nonlinear PDE (15.17). The results obtained for selected nodes x_i are given in Table 15.2 for two t values $t = 0.01$ and 0.1 along with the exact solution [10].

Next, we take $N = 7$ to find the solution of the PDE (15.7) using the DQM. Here, we get seven grid points $\{x_i, i = 1, 2, 3, 4, 5, 6, 7\}$. These are the roots of Legendre polynomial equation of order 7

$$P_7^*(x) = 3\,432x^7 - 12\,012x^6 + 166\,632x^5 - 11\,550x^4 + 4\,200x^3 - 756x^2 \\ + 56x - 1 = 0.$$

So, the nodes x_i are

$$x_1 = 0.025\,446\,04; \quad x_2 = 0.129\,234\,40; \quad x_3 = 0.297\,077\,42; \quad x_4 = 0.5;$$
$$x_5 = 0.702\,922\,57; \quad x_6 = 0.800\,765\,59; \quad x_7 = 0.974\,553\,95$$

Proceeding as above using Eqs. (15.7) and (15.8), we may find weighted coefficients $[a_{ij}]$ for $i = 1, 2, \ldots, 7$. Substituting 49 weighted coefficients $[a_{ij}]$ and 7 grid points x_i in Eq. (15.18) gives a nonlinear system of ODEs with constant coefficients. Then, the results obtained are presented in Table 15.3 after solving the system of nonlinear ODEs using the fourth-order Runge–Kutta method.

Table 15.2 Comparison of solutions of PDE (15.17) by the DQM for $N = 4$ with the exact solution.

t	x	DQM solution	Exact solution
0.01	x_1	2.65887716	2.00059430
	x_2	2.03204598	2.00320084
	x_3	1.91820603	2.00659969
	x_4	1.56030917	2.00920538
0.1	x_1	2.74196598	1.99696163
	x_2	2.03117563	2.02294138
	x_3	1.90734032	2.05681811
	x_4	1.52403620	2.08278936

Table 15.3 Comparison of solutions of PDE (15.17) by the DQM for $N = 7$ with the exact solution.

t	x	DQM solution	Exact solution
0.01	x_1	1.99986243	2.00015446
	x_2	2.00148324	2.00119231
	x_3	2.00387433	2.00287068
	x_4	2.00688137	2.00489984
	x_5	2.01417620	2.00692900
	x_6	2.01936977	2.00790739
	x_7	2.03492958	2.00964522
0.1	x_1	1.99859172	1.99257765
	x_2	2.01483750	2.00292203
	x_3	2.03886500	2.01965061
	x_4	2.06911125	2.03987550
	x_5	2.14410838	2.06010038
	x_6	2.19849403	2.06985220
	x_7	2.36865789	2.08717334

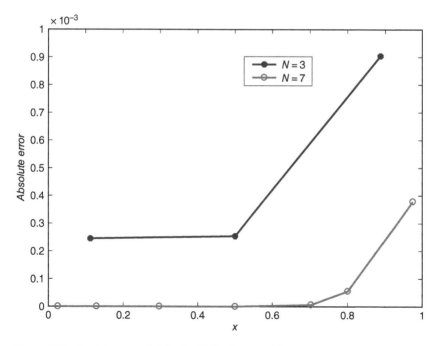

Figure 15.1 Absolute error plots for $N = 3, N = 7$ at $t = 0.01$.

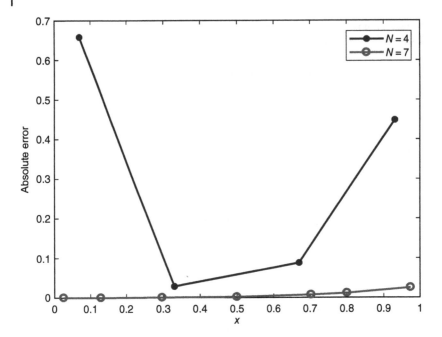

Figure 15.2 Absolute error plots for $N = 4, N = 7$ at $t = 0.01$.

From Tables 15.2 and 15.3, one may confirm that the results obtained are in good agreement with the exact solution. It may be observed that error in solution for $N = 4$ is more than that for $N = 7$.

Finally, error analysis has been done for both the problems discussed above in Examples 15.1 and 15.2. In the case of Example 15.1, if we consider $N=7$, the error between the solutions by the DQM and exact is very less. The corresponding error plots for $N=3$ and $N = 7$ at $t = 0.01$ are depicted in Figure 15.1. The similar type of behavior may be observed in the case of Example 15.2 also. Figure 15.2 represents the error plots for $N = 4$ and $N = 7$ at $t = 0.01$ for Example 15.2.

From Figures 15.1 and 15.2, one may conclude that the error is drastically decreasing as the N value increases from 3 and 4 to 7 as shown in Examples 15.1 and 15.2, respectively.

Exercise

1 Find the approximate solution of the solution of PDE $u_t(x, t) = x^2 - \frac{1}{4}u_x^2(x, t)$, subject to the initial condition $u(x, 0) = 0$ using the DQM with $N = 3, 5,$ and 7. Compare the approximate DQM solution in each case with the exact solution $u(x, t) = x^2 \tanh(t)$.

References

1 Bellman, R. and Casti, J. (1971). Differential quadrature and long-term integration. *Journal of Mathematical Analysis and Applications* 34: 235–238.

2 Bellman, R., Kashef, B.G., and Casti, J. (1972). Differential quadrature: a technique for the rapid solution of nonlinear partial differential equations. *Journal of Computational Physics* 10: 40–52.

3 Bellman, R., Kashef, B.G., Lee, E.S., and Vasudevan, R. (1975a). Solving hard problems by easy methods: differential and integral quadrature. *Computers and Mathematics with Applications* 1: 133–143.

4 Bellman, R., Kashef, B.G., Lee, E.S., and Vasudevan, R. (1975b). Differential quadrature and splines. *Computers and Mathematics with Applications* 1: 371–376.

5 Bert, C.W., Jang, S.K., and Striz, A.G. (1989). Nonlinear bending analysis of orthotropic rectangular plates by the method of differential quadrature. *Computational Mechanics* 5: 217–226.

6 Bert, C.W. and Kang, K.J. (1996). Stress analysis of closely-coiled helical springs using differential quadrature. *Mechanics Research Communications* 23 (6): 589–598.

7 Bert, C.W. and Malik, M. (1996). Differential quadrature method in computational mechanics. *Applied Mechanics Reviews* 49 (1): 1–28.

8 Shu, C. and Chen, W. (1999). On optimal selection of interior points for applying discretized boundary conditions in DQ vibration analysis of beams and plates. *Journal of Sound and Vibration* 222 (2): 239–257.

9 Shu, C. (1999). Application of differential quadrature method to simulate natural convection in a concentric annulus. *International Journal for Numerical Methods in Fluids* 30: 977–993.

10 Wazwaz, A.M. (2010). *Partial Differential Equations and Solitary Waves Theory*. Beijing: Springer Science & Business Media, Higher Education Press.

11 Ezz-Eldien, S.S., Doha, E.H., Baleanu, D., and Bhrawy, A.H. (2017). A numerical approach based on Legendre orthonormal polynomials for numerical solutions of fractional optimal control problems. *Journal of Vibration and Control* 23 (1): 16–30.

12 Bhrawy, A.H., Alofi, A.S., and Ezz-Eldien, S.S. (2011). A quadrature tau method for fractional differential equations with variable coefficients. *Applied Mathematics Letters* 24 (12): 2146–2152.

13 Debnath, L. (2012). *Nonlinear Partial Differential Equations for Scientists and Engineers*. New York: Springer Science & Business Media.

16

Wavelet Method

16.1 Introduction

Wavelet method has been proven to be an efficient tool in analyzing dynamic systems and differential equations arising in other science and engineering problems. A wave-like oscillation having amplitude beginning at zero that monotonically increases and then decreases back to zero is referred to as wavelet. Wavelets also serve as a tool in analysis of transient, nonstationary, and time-variate phenomena. A wavelet series is often represented in terms of complete square-integrable function or set of orthonormal basis functions. The broad classification of wavelet classes is considered as discrete, continuous, and multiresolution-based wavelets. Continuous wavelets are projected over continuous function space whereas discrete wavelets are often considered over discrete subset of upper half plane. While analyzing discrete wavelets, only a finite number of wavelet coefficients are taken into consideration which sometimes makes them numerically complex. In such cases, multiresolution-based wavelets are preferred.

In mid-1980s, orthonormal wavelets were initially studied by Meyer [1] over real line $\Re$. Gradually, compactly supported orthogonal wavelets viz. Daubechies wavelets [2] were constructed by Ingrid Daubechies. Daubechies wavelets are sufficiently smooth, possess orthogonality, and have compact support. A detailed study of wavelets and their applications may be found in Refs. [3–6]. Chen et al. [7] studied the computation of differential-integral equation using the wavelet–Galerkin method involving Daubechies wavelet and its derivatives or integrals. But, Daubechies wavelets are rather difficult in handling due to lack of explicit expression whereas Haar wavelets are computationally easier. As such, the next section discusses preliminaries related to Haar wavelet only.

Advanced Numerical and Semi-Analytical Methods for Differential Equations, First Edition.
Snehashish Chakraverty, Nisha Rani Mahato, Perumandla Karunakar, and Tharasi Dilleswar Rao.
© 2019 John Wiley & Sons, Inc. Published 2019 by John Wiley & Sons, Inc.

16.2   Haar Wavelet

In 1990, the concept of Haar wavelet [8] was introduced by Hungarian mathematician, Alfred Haar. Later, the pioneer work for system analysis was performed by Chen and Hsiao [9], who derived an operational matrix for integral of Haar function vector. Solutions of differential equations using Haar wavelets have been developed by various authors as given in Refs. [10–12] and the references mentioned therein. Further, a review of the Haar wavelet method for solving differential equations along with its advantage, applications, and comparative studies with other wavelets have been discussed in detail by Hariharan and Kannan [13].

Generally, the family of Haar wavelets [11, 13, 14] $h_i(x)$ for $x \in [0, 1]$ is considered as

$$h_i(x) = \begin{cases} 1, & x \in \left[\dfrac{k}{m}, \dfrac{k+0.5}{m}\right) \\ -1, & x \in \left[\dfrac{k+0.5}{m}, \dfrac{k+1}{m}\right) \\ 0, & \text{otherwise} \end{cases} \tag{16.1}$$

where the index i is calculated in terms of wavelet level m and translational parameter k using $i = m + k + 1$. Here, the resolution level of the wavelet is computed using integer $m = 2^j$ for $j = 0, 1, \ldots, J$ such that J denotes the maximal resolution level. Further, the translation parameter is computed using $k = 0, 1, \ldots, m - 1$. In this regard, the minimal and maximal values of index i are obtained as $i = 2$ and $i = 2M = 2^{J+1}$, respectively. The value of $h_i(x)$ at $i = 1$ may be assumed for $x \in [0, 1]$ as the scaling function given by

$$h_i(x) = \begin{cases} 1, & x \in [0, 1) \\ 0, & \text{otherwise} \end{cases}. \tag{16.2}$$

Let us now calculate the Haar wavelet functions using Eq. (16.1) over $0 \le x < 1$.

1. Initially, the values of M and j are obtained for maximal resolution level $J = 0$ as follows:

$$2M = 2^{J+1} \Rightarrow 2M = 2^{0+1} \Rightarrow M = 1$$

and
$$j = 0, 1, \ldots, J \Rightarrow j = 0.$$
Substitute $j = 0$ in $m = 2^j$ to obtain the value of $m = 1$. Then, the value of k is obtained using $k = 0, 1, \ldots, m - 1$ as $k = 0$. Index $i = 2$ is computed by substituting $m = 1$ and $k = 0$ in $i = m + k + 1$. Subsequently, the obtained values of i, k, and m are substituted in Eq. (16.1) to obtain Haar wavelet functions.

Further, using Eqs. (16.1) and (16.2) over $0 \leq x < 1$, the first two Haar wavelets are computed as

$$h_1(x) = \begin{cases} 1, & x \in [0,1) \\ 0, & \text{otherwise} \end{cases} \quad \text{and} \quad h_2(x) = \begin{cases} 1, & x \in \left[0, \frac{1}{2}\right) \\ -1, & x \in \left[\frac{1}{2}, 1\right) \\ 0, & \text{otherwise} \end{cases}.$$

2. Accordingly, the first four Haar wavelet functions may be generated for $J = 1$ as follows:

The values of M, j, m, k and i for maximal resolution level $J = 1$ are obtained using,

$$2M = 2^{J+1} \Rightarrow 2M = 2^{1+1} \Rightarrow M = 2,$$
$$j = 0, 1, \dots, J \Rightarrow j = 0, 1,$$
$$m = 2^j \Rightarrow m = 1, 2,$$
$$k = 0, 1, \dots, m-1 \Rightarrow k = 0, 1$$

and

$$i = m + k + 1 \Rightarrow i = 2, 3, 4,$$

respectively. By plugging $i = 2, 3, 4$, $k = 0, 1$, and $m = 1, 2$, the subsequent Haar wavelets are accordingly obtained. Then, using Eqs. (16.1) and (16.2) over $0 \leq x < 1$, the first four Haar wavelet functions are obtained as

$$h_1(x) = \begin{cases} 1, & x \in [0,1) \\ 0, & \text{otherwise} \end{cases}, \quad h_2(x) = \begin{cases} 1, & x \in \left[0, \frac{1}{2}\right) \\ -1, & x \in \left[\frac{1}{2}, 1\right) \\ 0, & \text{otherwise} \end{cases},$$

$$h_3(x) = \begin{cases} 1, & x \in \left[0, \frac{1}{4}\right) \\ -1, & x \in \left[\frac{1}{4}, \frac{1}{2}\right) \\ 0, & \text{otherwise} \end{cases} \quad \text{and} \quad h_4(x) = \begin{cases} 1, & x \in \left[\frac{1}{2}, \frac{3}{4}\right) \\ -1, & x \in \left[\frac{3}{4}, 1\right) \\ 0, & \text{otherwise} \end{cases}.$$

It is worth mentioning that the initial two Haar wavelets generated for $J = 1$ are equivalent to the Haar wavelets obtained for $J = 0$. Similarly, higher Haar wavelet functions may be obtained using Eqs. (16.1) and (16.2) for

$J > 1$. Further, the integrals of wavelets given by Eq. (16.1) may be calculated using [11, 14],

$$p_i(x) = \int_0^x h_i(x)dx = \begin{cases} x - \dfrac{k}{m}, & x \in \left[\dfrac{k}{m}, \dfrac{k+0.5}{m}\right) \\ \dfrac{k+1}{m} - x, & x \in \left[\dfrac{k+0.5}{m}, \dfrac{k+1}{m}\right) \\ 0, & \text{otherwise} \end{cases} \tag{16.3}$$

and

$$q_i(x) = \int_0^x p_i(x)dx = \begin{cases} 0, & x \in \left[0, \dfrac{k}{m}\right) \\ \dfrac{1}{2}\left(x - \dfrac{k}{m}\right)^2, & x \in \left[\dfrac{k}{m}, \dfrac{k+0.5}{m}\right) \\ \dfrac{1}{4m^2} - \dfrac{1}{2}\left(\dfrac{k+1}{m} - x\right)^2, & x \in \left[\dfrac{k+0.5}{m}, \dfrac{k+1}{m}\right) \\ \dfrac{1}{4m^2}, & x \in \left[\dfrac{k+1}{m}, 1\right) \end{cases} \tag{16.4}$$

As such, the first four Haar wavelets and their integrals using Eqs. (16.1)–(16.4) are obtained and incorporated into Table 16.1.

Accordingly, the respective plots of Haar wavelets and their integrals have been depicted in Table 16.2.

Wavelets serve as a very powerful tool for solving differential equations. In this context, only a basic idea of Haar wavelet is incorporated into this section. Further, just to have an overview of how to handle ordinary differential equation using Haar wavelets, a preliminary procedure is demonstrated. In this regard, Haar wavelet-based wavelet–collocation method [14] for solving ordinary differential equation has been discussed in the next section.

16.3 Wavelet–Collocation Method

Let us consider a second-order ordinary differential equation,

$$u'' + P(x)u' + Q(x)u = f(x) \tag{16.5}$$

subject to initial conditions $u(0)$ and $u'(0)$ over the domain Ω. In order to solve the differential equation (16.5), highest-order derivative is expressed in terms of linear combination of Haar wavelet functions [14] as

$$u''(x) = \sum_i c_i h_i(x) \tag{16.6}$$

Table 16.1 First four Haar wavelet functions and their integrals [14].

Haar functions, $h_i(x)$	Integral, $p_i(x)$	Integral, $q_i(x)$

$$h_1(x) = \begin{cases} 1, & x \in [0,1) \\ 0, & \text{otherwise} \end{cases}$$

$$p_1(x) = \begin{cases} x, & x \in [0,1) \\ 0, & \text{otherwise} \end{cases}$$

$$q_1(x) = \begin{cases} \dfrac{x^2}{2}, & x \in [0,1) \\ 0, & \text{otherwise} \end{cases}$$

$$h_2(x) = \begin{cases} 1, & x \in \left[0, \tfrac{1}{2}\right) \\ -1, & x \in \left[\tfrac{1}{2}, 1\right) \\ 0, & \text{otherwise} \end{cases}$$

$$p_2(x) = \begin{cases} x, & x \in \left[0, \tfrac{1}{2}\right) \\ 1-x, & x \in \left[\tfrac{1}{2}, 1\right) \\ 0, & \text{otherwise} \end{cases}$$

$$q_2(x) = \begin{cases} \tfrac{1}{2}x^2, & x \in \left[0, \tfrac{1}{2}\right) \\ \tfrac{1}{4} - \tfrac{1}{2}(1-x)^2, & x \in \left[\tfrac{1}{2}, 1\right) \\ \tfrac{1}{4}, & x \in [1,1) \end{cases}$$

$$h_3(x) = \begin{cases} 1, & x \in \left[0, \tfrac{1}{4}\right) \\ -1, & x \in \left[\tfrac{1}{4}, \tfrac{1}{2}\right) \\ 0, & \text{otherwise} \end{cases}$$

$$p_3(x) = \begin{cases} x, & x \in \left[0, \tfrac{1}{4}\right) \\ \tfrac{1}{2} - x, & x \in \left[\tfrac{1}{4}, \tfrac{1}{2}\right) \\ 0, & \text{otherwise} \end{cases}$$

$$q_3(x) = \begin{cases} \tfrac{1}{2}x^2, & x \in \left[0, \tfrac{1}{4}\right) \\ \tfrac{1}{16} - \tfrac{1}{2}\left(\tfrac{1}{4} - x\right)^2, & x \in \left[\tfrac{1}{4}, \tfrac{1}{2}\right) \\ \tfrac{1}{16}, & x \in \left[\tfrac{1}{2}, 1\right) \end{cases}$$

$$h_4(x) = \begin{cases} 1, & x \in \left[\tfrac{1}{2}, \tfrac{3}{4}\right) \\ -1, & x \in \left[\tfrac{3}{4}, 1\right) \\ 0, & \text{otherwise} \end{cases}$$

$$p_4(x) = \begin{cases} x - \tfrac{1}{2}, & x \in \left[\tfrac{1}{2}, \tfrac{3}{4}\right) \\ 1 - x, & x \in \left[\tfrac{3}{4}, 1\right) \\ 0, & \text{otherwise} \end{cases}$$

$$q_4(x) = \begin{cases} 0, & x \in \left[0, \tfrac{1}{2}\right) \\ \tfrac{1}{2}\left(x - \tfrac{1}{2}\right)^2, & x \in \left[\tfrac{1}{2}, \tfrac{3}{4}\right) \\ \tfrac{1}{16} - \tfrac{1}{2}(1-x)^2, & x \in \left[\tfrac{3}{4}, 1\right) \\ \tfrac{1}{16}, & x \in [1,1) \end{cases}$$

Table 16.2 Plots of first four Haar wavelets and their integrals.

i	Haar functions, $h_i(x)$	Integral, $p_i(x)$	Integral, $q_i(x)$
1			
2			
3			
4			

where c_i for $i = 1, 2, \ldots$ are the unknown wavelet coefficients yet to be determined. Accordingly, the lower derivative $u'(x)$ and the solution $u(x)$ are expressed in terms of integrals of Haar wavelet functions as

$$u'(x) = \sum_i c_i p_i(x) + u'(0) \tag{16.7}$$

and

$$u(x) = \sum_i c_i q_i(x) + xu'(0) + u(0). \tag{16.8}$$

Now, substitute Eqs. (16.6)–(16.8) in the governing differential equation (16.5) to obtain

$$\sum_i c_i(h_i(x) + P(x)p_i(x) + Q(x)q_i(x)) + (P(x) + xQ(x))u'(0) + u(0) = f(x).$$

$$\tag{16.9}$$

Then, the collocation points for solving Eq. (16.5) using the Haar wavelet–collocation method may be obtained using [10, 11, 14]

$$x_r = \frac{(r - 0.5)}{2M} \tag{16.10}$$

where $r = 1, 2, \ldots, 2M$. Generally, the technique of Haar wavelet–collocation method transforms the differential equations to system of algebraic equations. In this regard, the system of equations in terms of unknown coefficients c_i is obtained for the collocation points x_r given by Eq. (16.10). Finally, the approximate solution is computed using Eq. (16.8) as

$$\hat{u}(x) = \sum_{i=1}^{2M} c_i q_i(x) + xu'(0) + u(0) \tag{16.11}$$

The effectiveness of the above-mentioned procedure is illustrated using an example problem.

Example 16.1   Solve an initial value problem,

$$u'' - 2u' + u = x, \tag{16.12}$$

with respect to initial conditions $u(0) = 1$ and $u'(0) = 0$ using the wavelet–collocation method.

Solution   Using Eq. (16.9), the differential equation (16.12) reduces to

$$\sum_{i=1}^{2M} c_i[h_i(x) - 2p_i(x) + q_i(x)] = x - 1. \tag{16.13}$$

The collocation points are obtained utilizing Eq. (16.10) for $M = 2$ as $x_1 = 1/8$, $x_2 = 3/8$, $x_3 = 5/8$, and $x_4 = 7/8$. Accordingly, Eq. (16.13) reduces to system of equation given by

$$C(H - 2P + Q) = X - 1 \tag{16.14}$$

where $H = \begin{bmatrix} 1 & 1 & 1 & 1 \\ 1 & 1 & -1 & -1 \\ 1 & -1 & 0 & 0 \\ 0 & 0 & 1 & -1 \end{bmatrix}$, $P = \frac{1}{8}\begin{bmatrix} 1 & 3 & 5 & 7 \\ 1 & 3 & 3 & 1 \\ 1 & 1 & 0 & 0 \\ 0 & 0 & 1 & 1 \end{bmatrix}$, $Q = \frac{1}{128}\begin{bmatrix} 1 & 9 & 25 & 49 \\ 1 & 9 & 23 & 31 \\ 1 & 7 & 8 & 8 \\ 0 & 0 & 1 & 7 \end{bmatrix}$,

$C = \{c_1, c_2, c_3, c_4\}$, and $X = \{x_1, x_2, x_3, x_4\}$. Further, Eq. (16.14) gets converted to

$$\frac{1}{128}\{c_1, c_2, c_3, c_4\}\begin{bmatrix} 97 & 41 & -7 & -47 \\ 97 & 41 & -201 & -129 \\ 97 & -153 & 8 & 8 \\ 0 & 0 & 97 & -153 \end{bmatrix} = -\left\{\frac{7}{8}, \frac{5}{8}, \frac{3}{8}, \frac{1}{8}\right\}. \tag{16.15}$$

On solving Eq. (16.15), the wavelet coefficients are obtained as $c_1 = -1.7662$, $c_2 = 0.4432$, $c_3 = 0.1683$, and $c_4 = 0.2822$.

Table 16.3 Comparison of solution obtained using the Haar wavelet with the exact solution.

x	Haar wavelet–collocation solution, $\hat{u}(x)$	Exact solution, $u(x)$
$\dfrac{1}{8}$	0.9910	0.9919
$\dfrac{3}{8}$	0.9162	0.9200
$\dfrac{5}{8}$	0.7474	0.7568
$\dfrac{7}{8}$	0.4572	0.4761

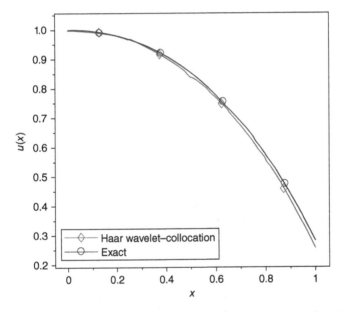

Figure 16.1 Plots of solution obtained using the Haar wavelet–collocation method and exact solution.

The approximate solution $\hat{u}(x)$ is obtained by substituting the above coefficients in Eq. (16.11). Then, the results at the collocation points are compared in Table 16.3 and Figure 16.1 with the exact solution $u(x) = x - e^x + 2$ of the given differential equation (16.12).

It is worth mentioning that better approximation of solution may be obtained using the Haar wavelet–collocation technique by considering more collocation points with respect to higher Haar wavelet functions for $i > 4$ in Eq. (16.1).

In case of partial differential equations, the two-dimensional Haar wavelet transform may be used for solving governing equation using the wavelet–collocation method [11]. Various other methods in this regard may be found in Refs. [15, 16].

Exercise

1 Solve the linear differential equation $\frac{d^2u}{dt^2} + 2\frac{du}{dt} - 5u = \cos(2t)$ subject to initial condition $u(0) = 0$ and $\frac{du}{dt}(0) = 1$.

2 Evaluate the Duffing equation $\frac{d^2u}{dt^2} + 0.25\frac{du}{dt} - u + u^3 = 0.2\cos(2t)$ with respect to initial conditions $u(0) = 0$ and $\frac{du}{dt}(0) = 1$ for $t \in [0, 1]$.

References

1 Meyer, Y. (1989). Orthonormal wavelets. In: *Wavelets* (ed. J.-M. Combes, A. Grossmann and P. Tchamitchian), 21–37. Berlin, Heidelberg: Springer.

2 Daubechies, I. (1988). Orthonormal bases of compactly supported wavelets. *Communications on Pure and Applied Mathematics* 41 (7): 909–996.

3 Daubechies, I. (1992). *Ten Lectures on Wavelets*, vol. 61. Philadelphia, PA: SIAM.

4 Benedetto, J.J. (1993). *Wavelets: Mathematics and Applications*, vol. 13. Boca Raton, FL: CRC Press.

5 Strang, G. and Nguyen, T. (1996). *Wavelets and Filter Banks*. Wellesley, MA: SIAM, Wellesley-Cambridge Press.

6 Chui, C.K. (2016). *An Introduction to Wavelets*. San Diego: Academic Press.

7 Chen, M.Q., Hwang, C., and Shih, Y.P. (1996). The computation of wavelet-Galerkin approximation on a bounded interval. *International Journal for Numerical Methods in Engineering* 39 (17): 2921–2944.

8 Haar, A. (1910). Zur theorie der orthogonalen funktionensysteme. *Mathematische Annalen* 69 (3): 331–371.

9 Chen, C.F. and Hsiao, C.H. (1997). Haar wavelet method for solving lumped and distributed-parameter systems. *IEE Proceedings-Control Theory and Applications* 144 (1): 87–94.

10 Lepik, Ü. (2005). Numerical solution of differential equations using Haar wavelets. *Mathematics and Computers in Simulation* 68 (2): 127–143.

11 Lepik, Ü. (2007). Numerical solution of evolution equations by the Haar wavelet method. *Applied Mathematics and Computation* 185 (1): 695–704.

12 Shi, Z., Deng, L.Y. and Chen, Q.J., 2007, Numerical solution of differential equations by using Haar wavelets. *Proceedings of the International Conference on Wavelet Analysis and Pattern Recognition, 2007. ICWAPR'07, Beijing*, (2–4 November 2007), Vol. 3, 1039–1044. IEEE.

13 Hariharan, G. and Kannan, K. (2010). Haar wavelet method for solving some nonlinear parabolic equations. *Journal of Mathematical Chemistry* 48 (4): 1044–1061.

14 Nayak, S. and Chakraverty, S. (2016). Interval wavelet method in solving diffusion equations, generalized and hybrid set structures and applications for soft computing. In: *Handbook of Research on Generalized and Hybrid Set Structures and Applications for Soft Computing* (ed. S.J. John), 457–472. Hershey, PA: IGI Publication.

15 Comincioli, V., Naldi, G., and Scapolla, T. (2000). A wavelet-based method for numerical solution of nonlinear evolution equations. *Applied Numerical Mathematics* 33 (1–4): 291–297.

16 Lazaar, S., Ponenti, P.J., Liandrat, J., and Tchamitchian, P. (1994). Wavelet algorithms for numerical resolution of partial differential equations. *Computer Methods in Applied Mechanics and Engineering* 116 (1–4): 309–314.

17

Hybrid Methods

17.1   Introduction

Laplace transform method, Adomian decomposition method (ADM), and homotopy perturbation method (HPM) have been implemented for solving various types of differential equations in Chapters 2, 11, and 12, respectively. This chapter deals with the methods that combine more than one method. These have been termed as hybrid methods. Two main such methods, which are getting more attention of research, are homotopy perturbation transform method (HPTM) and Laplace Adomian decomposition method (LADM). Although the Laplace transform method is an effective method for solving ordinary and partial differential equations (PDEs), a notable difficulty with this method is about handling nonlinear terms if appearing in the differential equations. This difficulty may be overcome by combining Laplace transform (LT) with the HPM and ADM.

First, we briefly describe the HPTM in the following section.

17.2   Homotopy Perturbation Transform Method

Here, the Laplace transform method and HPM are combined to have a method called the HPTM for solving nonlinear differential equations. It is also called as Laplace homotopy perturbation method (LHPM).

Let us consider a general nonlinear PDE with source term $g(x, t)$ to illustrate the basic idea of HPTM as below [1–3]

$$Du(x, t) + Ru(x, t) + Nu(x, t) = g(x, t) \tag{17.1}$$

subject to initial conditions

$$u(x, 0) = h(x), \quad \frac{\partial}{\partial t} u(x, 0) = f(x) \tag{17.2}$$

Advanced Numerical and Semi-Analytical Methods for Differential Equations, First Edition.
Snehashish Chakraverty, Nisha Rani Mahato, Perumandla Karunakar, and Tharasi Dilleswar Rao.
© 2019 John Wiley & Sons, Inc. Published 2019 by John Wiley & Sons, Inc.

where D is the linear differential operator $D = \frac{\partial^2}{\partial t^2}$ (or $\frac{\partial}{\partial t}$), R is the linear differential operator whose order is less than that of D, and N is the nonlinear differential operator.

The HPTM methodology consists of mainly two steps. The first step is applying LT on both sides of Eq. (17.1) and the second step is applying the HPM where decomposition of the nonlinear term is done using He's polynomials.

First, by operating LT on both sides of Eq. (17.1), we obtain

$$\mathcal{L}[Du(x, t)] = -\mathcal{L}[Ru(x, t)] - \mathcal{L}[Nu(x, t)] + \mathcal{L}[g(x, t)]$$

Assuming that D is a second-order differential operator and using differentiation property of LT we get

$$s^2 \mathcal{L}[u(x, t)] - sh(x) - f(x) = -\mathcal{L}[Ru(x, t)] - \mathcal{L}[Nu(x, t)] + \mathcal{L}[g(x, t)]$$

$$\mathcal{L}[u(x, t)] = \frac{h(x)}{s} + \frac{f(x)}{s^2} + \frac{1}{s^2}\mathcal{L}[g(x, t)] - \frac{1}{s^2}\mathcal{L}[Ru(x, t)]$$

$$- \frac{1}{s^2}\mathcal{L}[Nu(x, t)] \tag{17.3}$$

By applying inverse LT on both sides of Eq. (17.3), we have

$$u(x, t) = G(x, t) - \mathcal{L}^{-1}\left[\frac{1}{s^2}\mathcal{L}[Ru(x, t)] + \frac{1}{s^2}\mathcal{L}[Nu(x, t)]\right] \tag{17.4}$$

where $G(x, t)$ is the term arising from first three terms on right-hand side of Eq. (17.3).

Next, to apply HPM (Chapter 12), first we need to assume solution as series that contains embedding parameter $p \in [0, 1]$ as

$$u(x, t) = \sum_{n=0}^{\infty} p^n u_n(x, t) \tag{17.5}$$

and the nonlinear term may be decomposed using He's polynomials as

$$Nu(x, t) = \sum_{n=0}^{\infty} p^n H_n(u) \tag{17.6}$$

where $H_n(u)$ represents the He's polynomials [1, 2] which are defined as below

$$H_n(u_0, u_1, \ldots, u_n) = \frac{1}{n!}\left[\frac{\partial^n}{\partial p^n}N\left(\sum_{i=0}^{\infty} p^i u_i\right)\right]_{p=0}, \quad n = 0, 1, 2, 3, \ldots \tag{17.7}$$

Interested readers are suggested to see Ref. [4] for the detailed derivation of He' polynomials (Eq. (17.7)) using the concept of HPM. By plugging Eqs. (17.5) and (17.6) in Eq. (17.4) and combining LT with the HPM, one may obtain the following expression:

$$\sum_{n=0}^{\infty} p^n u_n(x,t) = G(x,t)$$

$$- p\left(\mathcal{L}^{-1}\left[\frac{1}{s^2}\mathcal{L}\left[R\sum_{n=0}^{\infty} p^n u_n(x,t) \right] + \frac{1}{s^2}\mathcal{L}\left[\sum_{n=0}^{\infty} p^n H_n(u) \right] \right] \right). \quad (17.8)$$

By comparing the coefficients of like powers of "p" on both sides of Eq. (17.8) we may obtain the following successive approximations:

$$p^0 : u_0(x,t) = G(x,t)$$

$$p^1 : u_1(x,t) = -\mathcal{L}^{-1}\left[\frac{1}{s^2}\mathcal{L}[Ru_0(x,t)] + \frac{1}{s^2}\mathcal{L}[H_0(u)] \right],$$

$$p^2 : u_2(x,t) = -\mathcal{L}^{-1}\left[\frac{1}{s^2}\mathcal{L}[Ru_1(x,t)] + \frac{1}{s^2}\mathcal{L}[H_1(u)] \right],$$

$$p^3 : u_3(x,t) = -\mathcal{L}^{-1}\left[\frac{1}{s^2}\mathcal{L}[Ru_2(x,t)] + \frac{1}{s^2}\mathcal{L}[H_2(u)] \right],$$

$$\vdots$$

$$p^n : u_n(x,t) = -\mathcal{L}^{-1}\left[\frac{1}{s^2}\mathcal{L}[Ru_{n-1}(x,t)] + \frac{1}{s^2}\mathcal{L}[H_{n-1}(u)] \right],$$

$$\vdots$$

Finally, the solution of the differential equation (17.1) may be obtained as below

$$u(x,t) = \lim_{p \to 1} u_n(x,t) = u_0(x,t) + u_1(x,t) + u_2(x,t) + \cdots \quad (17.9)$$

Next, we solve two test problems to demonstrate the present method.

Example 17.1 Consider advection equation from Ref. [5] as

$$u_t + uu_x = x \quad (17.10)$$

with initial condition

$$u(x,0) = 2 \quad (17.11)$$

Solution Applying LT on both sides of Eq. (17.10) subject to initial condition (17.11), we get

$$\mathcal{L}[u(x,t)] = \frac{2}{s} + \frac{1}{s}\mathcal{L}[x] - \frac{1}{s}\mathcal{L}[uu_x]$$

$$\mathcal{L}[u(x,t)] = \frac{2}{s} + \frac{1}{s}\left(\frac{x}{s}\right) - \frac{1}{s}\mathcal{L}[uu_x] \quad (17.12)$$

Taking inverse LT on both sides of Eq. (17.12) gives

$$u(x,t) = 2 + xt - \mathcal{L}^{-1}\left[\frac{1}{s}\mathcal{L}[uu_x] \right]$$

By using the HPM procedure, we get

$$\sum_{n=0}^{\infty} p^n u_n(x,t) = 2 + xt - p\left(\mathcal{L}^{-1}\left[\frac{1}{s}\mathcal{L}\left[\sum_{n=0}^{\infty} p^n H_n(u)\right]\right]\right) \tag{17.13}$$

The first few He's polynomials $H_n(u)$ [2] for the nonlinear term uu_x are

$$\left.\begin{aligned}
H_0(u) &= u_0 \frac{\partial u_0}{\partial x} \\
H_1(u) &= u_0 \frac{\partial u_1}{\partial x} + u_1 \frac{\partial u_0}{\partial x} \\
H_2(u) &= u_0 \frac{\partial u_2}{\partial x} + u_1 \frac{\partial u_1}{\partial x} + u_2 \frac{\partial u_0}{\partial x} \\
&\vdots
\end{aligned}\right\} \tag{17.14}$$

Comparing the coefficients of like powers of p on both sides of Eq. (17.13), we get

$$p^0 : u_0(x,t) = 2 + xt,$$

$$\begin{aligned}
p^1 : u_1(x,t) &= -\mathcal{L}^{-1}\left[\frac{1}{s}\mathcal{L}[H_0(u)]\right] \\
&= -\mathcal{L}^{-1}\left[\frac{1}{s}\mathcal{L}\left[u_0 \frac{\partial u_0}{\partial x}\right]\right] \\
&= -\mathcal{L}^{-1}\left[\frac{1}{s}\mathcal{L}\left[(2+xt)\frac{\partial}{\partial x}(2+xt)\right]\right] \\
&= -\mathcal{L}^{-1}\left[\frac{1}{s}\mathcal{L}[2t + xt^2]\right] \\
&= -\mathcal{L}^{-1}\left[\frac{2}{s^3} + \frac{2x}{s^4}\right] \\
&= -t^2 - \frac{xt^3}{3},
\end{aligned}$$

$$\begin{aligned}
p^2 : u_2(x,t) &= -\mathcal{L}^{-1}\left[\frac{1}{s}\mathcal{L}[H_1(u)]\right] \\
&= -\mathcal{L}^{-1}\left[\frac{1}{s}\mathcal{L}\left[u_0 \frac{\partial u_1}{\partial x} + u_1 \frac{\partial u_0}{\partial x}\right]\right] \\
&= \frac{15}{12}t^4 + \frac{2}{15}xt^5,
\end{aligned}$$

$$\vdots$$

The solution of the advection equation (17.10), as $p \to 1$ may be obtained as

$$u = 2 + xt - t^2 - \frac{1}{3}xt^3 + \frac{5}{12}t^4 + \frac{2}{15}xt^5 + \cdots$$

$$\Rightarrow u = 2\left(1 - \frac{1}{2}t^2 + \frac{5}{24}t^4 - \cdots\right) + x\left(t - \frac{1}{3}t^3 + \frac{2}{15}t^5 - \cdots\right)$$

$$\Rightarrow u = 2 \operatorname{sech} t + x \tanh t$$

which is the exact solution [5] of Eq. (17.10).

Example 17.2   Now, we apply the HPTM to Burger's equation [6]

$$\frac{\partial u}{\partial t} - \frac{\partial^2 u}{\partial x^2} + u\frac{\partial u}{\partial x} = 0, \tag{17.15}$$

subject to initial conditions

$$u(x, 0) = 2x. \tag{17.16}$$

Solution   The exact solution [6] of Eq. (17.15) with initial condition (17.16) is

$$u(x, t) = \frac{2x}{1 + 2t}. \tag{17.17}$$

By applying LT on both sides of Eq. (17.15) subject to initial condition (17.16), we get

$$\mathcal{L}[u(x, t)] = \frac{2x}{s} + \frac{1}{s}\mathcal{L}\left[\frac{\partial^2 u}{\partial x^2} - u\frac{\partial u}{\partial x}\right]. \tag{17.18}$$

Taking inverse LT on both sides of Eq. (17.18) gives

$$u(x, t) = 2x + \mathcal{L}^{-1}\left[\frac{1}{s}\mathcal{L}\left[\frac{\partial^2 u}{\partial x^2} - u\frac{\partial u}{\partial x}\right]\right]. \tag{17.19}$$

By applying the HPM to Eq. (17.19), we get

$$\sum_{n=0}^{\infty} p^n u_n(x, t) = 2x + p\left(\mathcal{L}^{-1}\left[\frac{1}{s}\mathcal{L}\left[\frac{\partial^2}{\partial x^2}\sum_{n=0}^{\infty} p^n u_n(x, t)\right] - \sum_{n=0}^{\infty} p^n H_n(u)\right]\right) \tag{17.20}$$

By comparing the coefficients of like powers of p on both sides of Eq. (17.20) and making use of He's polynomials (17.14), we obtain

$$p^0 : u_0(x, t) = 2x,$$

$$p^1 : u_1(x, t) = \mathcal{L}^{-1}\left[\frac{1}{s}\mathcal{L}\left[\frac{\partial^2 u_0}{\partial x^2} - H_0(u)\right]\right]$$

$$= \mathcal{L}^{-1}\left[\frac{1}{s}\mathcal{L}\left[\frac{\partial^2 u_0}{\partial x^2} - u_0\frac{\partial u_0}{\partial x}\right]\right]$$

$$= \mathcal{L}^{-1}\left[\frac{1}{s}\mathcal{L}[0 - 2x(2)]\right]$$

$$= \mathcal{L}^{-1}\left[\frac{-4x}{s^2}\right]$$

$$= -4xt$$

$$p^2 : u_2(x, t) = \mathcal{L}^{-1}\left[\frac{1}{s}\mathcal{L}\left[\frac{\partial^2 u_1}{\partial x^2} - H_1(u)\right]\right]$$

$$= \mathcal{L}^{-1}\left[\frac{1}{s}\mathcal{L}[0 - (-16xt)]\right]$$

$$= \mathcal{L}^{-1}\left[\frac{16x}{s^3}\right]$$

$$= 8xt^2.$$

Similarly,

$$p^3 : u_3(x, t) = -16xt^3,$$
$$p^4 : u_4(x, t) = 32xt^5,$$
$$p^5 : u_5(x, t) = -64xt^6.$$

Thus, the solution of the Burger's equation (17.15) with the given initial condition is

$$u(x, t) = 2x - 4xt + 8xt^2 - 16xt^3 + 32xt^4 - 64xt^5 + \cdots$$
$$= 2x(1 - 2t + 4t^2 - 8t^3 + 16t^4 - 32t^5 + \cdots)$$

The compact form solution of the above is $u(x, t) = \frac{2x}{1+2t}$ which is the same as the exact solution (17.17).

17.3   Laplace Adomian Decomposition Method

Now, we briefly illustrate the LADM [7, 8]. This method combines LT with the ADM. In this method nonlinear terms of the differential equations are handled by Adomian polynomials.

Let us consider nonlinear nonhomogeneous PDE (17.1) with same conditions as in PDE (17.2) for the explanation of the LADM.

Just like HPTM we apply LT on both sides of PDE (17.1) and using differentiation property of LT we obtain

$$s^2 \mathcal{L}[u(x,t)] - sh(x) - f(x) = \mathcal{L}[g(x,t)] - \mathcal{L}[Ru(x,t)] - \mathcal{L}[Nu(x,t)]$$
(17.21)

Inverse LT on both sides of PDE (17.21) results in

$$u(x,t) = G(x,t) - \mathcal{L}^{-1}\left[\frac{1}{s^2}\mathcal{L}[Ru(x,t)] + \frac{1}{s^2}\mathcal{L}[Nu(x,t)]\right],$$
(17.22)

where $G(x,t)$ is the term arising from first three terms on right-hand side of PDE (17.21).

In the LADM we assume a series solution for the PDE as below

$$u(x,t) = \sum_{n=0}^{\infty} u_n(x,t)$$
(17.23)

Here, the nonlinear term is written in terms of Adomian polynomials as

$$Nu(x,t) = \sum_{n=0}^{\infty} A_n$$
(17.24)

where A_n represents the Adomian polynomials [9] and these are defined as

$$A_n = \frac{1}{n!}\left[\frac{d^n}{dq^n}N\left(\sum_{i=0}^{\infty}q^i u_i\right)\right]_{q=0}, \quad n = 0,1,2,3,\ldots$$
(17.25)

By plugging Eqs. (17.23) and (17.24) in Eq. (17.22), we obtain

$$\sum_{n=0}^{\infty} u_n(x,t) = G(x,t) - \mathcal{L}^{-1}\left[\frac{1}{s^2}\mathcal{L}\left[R\sum_{n=0}^{\infty}u_n(x,t)\right] + \frac{1}{s^2}\mathcal{L}\left[\sum_{n=0}^{\infty}A_n\right]\right]$$
(17.26)

An iterative algorithm may be obtained by matching both sides of Eq. (17.26) as below

$$u_0(x,t) = G(x,t)$$
(17.27)

$$u_1(x,t) = -\mathcal{L}^{-1}\left[\frac{1}{s^2}\mathcal{L}[Ru_0(x,t)] + \frac{1}{s^2}\mathcal{L}[A_0]\right],$$

$$u_2(x,t) = -\mathcal{L}^{-1}\left[\frac{1}{s^2}\mathcal{L}[Ru_1(x,t)] + \frac{1}{s^2}\mathcal{L}[A_1]\right],$$

$$u_3(x,t) = -\mathcal{L}^{-1}\left[\frac{1}{s^2}\mathcal{L}[Ru_2(x,t)] + \frac{1}{s^2}\mathcal{L}[A_2]\right],$$

$$\vdots$$

$$u_n(x,t) = -\mathcal{L}^{-1}\left[\frac{1}{s^2}\mathcal{L}[Ru_{n-1}(x,t)] + \frac{1}{s^2}\mathcal{L}[A_{n-1}]\right].$$
(17.28)

Solution of the differential equation (17.1) using the LADM is

$$u(x, t) = u_0(x, t) + u_1(x, t) + u_2(x, t) + \cdots \tag{17.29}$$

Next, we apply LADM to a nonlinear PDE to make the readers familiar with the present method.

Example 17.3 Let us consider the Cauchy problem from Ref. [10]

$$4u_t - u_x^2 = 4x^2 \tag{17.30}$$

subject to initial conditions

$$u(x, 0) = 0. \tag{17.31}$$

Solution The exact solution [10] of the above PDE is

$$u(x, t) = x^2 \tan t. \tag{17.32}$$

Rewrite Eq. (17.30) as

$$u_t = x^2 + \frac{1}{4}u_x^2 \tag{17.33}$$

Using LT to Eq. (17.33), we get

$$\mathcal{L}[u_t(x, t)] = \frac{x^2}{s} + \frac{1}{4}\mathcal{L}[u_x^2]$$

$$s\mathcal{L}[u(x, t)] - u(x, 0) = \frac{x^2}{s} + \frac{1}{4}\mathcal{L}[u_x^2]$$

$$\Rightarrow \mathcal{L}[u(x, t)] = \frac{x^2}{s^2} + \frac{1}{4s}\mathcal{L}[u_x^2].$$

Taking inverse LT on both sides, we get

$$u(x, t) = x^2 t + \frac{1}{4}\mathcal{L}^{-1}\left[\frac{1}{s}\mathcal{L}[u_x^2]\right]. \tag{17.34}$$

According to the LADM, assuming a series solution for the PDE as Eq. (17.23) and by replacing the nonlinear term by Eq. (17.24), we obtain

$$\sum_{n=0}^{\infty} u_n(x, t) = x^2 t + \frac{1}{4}\mathcal{L}^{-1}\left[\frac{1}{s}\mathcal{L}\left[\sum_{n=0}^{\infty} A_n\right]^2\right] \tag{17.35}$$

where the Adomian polynomials A_n for the nonlinear term u_x^2 are given by [8]

$$
\left.
\begin{aligned}
A_0 &= u_{0x}^2 \\
A_1 &= 2u_{0x}u_{1x} \\
A_2 &= 2u_{0x}u_{2x} + u_{1x}^2 \\
A_3 &= 2u_{0x}u_{3x} + 2u_{1x}u_{2x} \\
A_4 &= 2u_{0x}u_{4x} + 2u_{1x}u_{4x} + u_{2x}^2 \\
&\vdots
\end{aligned}
\right\}
\tag{17.36}
$$

Next, successive approximations to the solution of Eq. (17.30) may be found using Eqs. (17.27) and (17.28) as

$$
u_0(x, t) = x^2 t
$$

$$
\begin{aligned}
u_1(x, t) &= \frac{1}{4}\mathcal{L}^{-1}\left[\frac{1}{s}\mathcal{L}[A_0]\right] \\
&= \frac{1}{4}\mathcal{L}^{-1}\left[\frac{1}{s}\mathcal{L}[u_{0x}^2]\right] \\
&= \frac{1}{4}\mathcal{L}^{-1}\left[\frac{1}{s}\mathcal{L}[(2xt)^2]\right] \\
&= \frac{1}{4}\mathcal{L}^{-1}\left[\frac{1}{s}4x^2\mathcal{L}[t^2]\right] \\
&= x^2\mathcal{L}^{-1}\left[\frac{1}{s}\mathcal{L}[t^2]\right] \\
&= x^2\mathcal{L}^{-1}\left[\frac{2}{s^4}\right] \\
&= x^2\frac{t^3}{3}
\end{aligned}
$$

$$
\begin{aligned}
u_2(x, t) &= \frac{1}{4}\mathcal{L}^{-1}\left[\frac{1}{s}\mathcal{L}[A_1]\right] \\
&= \frac{1}{4}\mathcal{L}^{-1}\left[\frac{1}{s}\mathcal{L}[2u_{0x}u_{1x}]\right] \\
&= \frac{1}{4}\mathcal{L}^{-1}\left[\frac{1}{s}\mathcal{L}\left[2(2xt)\left(2x\frac{t^3}{3}\right)\right]\right] \\
&= \frac{2}{3}x^2\mathcal{L}^{-1}\left[\frac{1}{s}\frac{4!}{s^5}\right] \\
&= \frac{2\times 4!}{3}x^2\mathcal{L}^{-1}\left[\frac{1}{s^6}\right] \\
&= x^2\frac{2}{15}t^5
\end{aligned}
$$

$$
u_3(x, t) = x^2\frac{17}{315}t^7
$$

$$
u_4(x, t) = x^2\frac{62}{2835}t^9
$$

$$\vdots$$

The solution of Eq. (17.30) is then written as

$$u(x, t) = x^2 t + x^2 \frac{2}{15} t^5 + x^2 \frac{17}{315} t^7 + x^2 \frac{62}{2835} t^9 + \cdots$$

$$= x^2 \left(t + \frac{2}{15} t^5 + \frac{17}{315} t^7 + x \frac{62}{2835} t^9 + \cdots \right)$$

This gives $u(x, t) = x^2 \tan t$ which is the exact solution as given in Eq. (17.32). Here, only two hybrid methods are addressed. In the similar fashion, readers may try to combine two or more methods for possible efficient solution.

Exercise

1 Use the HPTM to solve $u_t + u u_x = 0$ subject to initial condition $u(x, 0) = x$.

2 Use the LADM for finding the solution of ordinary nonlinear differential equation $u_t = u_{xx} + u(1 - u)$ subject to initial condition $u(x, 0) = 1$.

References

1 Madani, M., Fathizadeh, M., Khan, Y., and Yildirim, A. (2011). On the coupling of the homotopy perturbation method and Laplace transformation. *Mathematical and Computer Modelling* 53 (9–10): 1937–1945.

2 Khan, Y. and Wu, Q. (2011). Homotopy perturbation transform method for nonlinear equations using He's polynomials. *Computers and Mathematics with Applications* 61 (8): 1963–1967.

3 Aminikhah, H. (2012). The combined Laplace transform and new homotopy perturbation methods for stiff systems of ODEs. *Applied Mathematical Modelling* 36 (8): 3638–3644.

4 Ghorbani, A. (2009). Beyond Adomian polynomials: He polynomials. *Chaos, Solitons, and Fractals* 39 (3): 1486–1492.

5 Wazwaz, A.M. (2010). *Partial Differential Equations and Solitary Waves Theory*. Beijing: Springer Science & Business Media, Higher Education Press.

6 Gorguis, A. (2006). A comparison between Cole–Hopf transformation and the decomposition method for solving Burgers' equations. *Applied Mathematics and Computation* 173 (1): 126–136.

7 Jafari, H., Khalique, C.M., and Nazari, M. (2011). Application of the Laplace decomposition method for solving linear and nonlinear fractional diffusion–wave equations. *Applied Mathematics Letters* 24 (11): 1799–1805.

8 Khan, M., Hussain, M., Jafari, H., and Khan, Y. (2010). Application of Laplace decomposition method to solve nonlinear coupled partial differential equations. *World Applied Sciences Journal* 9 (1): 13–19.

9 Wazwaz, A.M. (2000). A new algorithm for calculating Adomian polynomials for nonlinear operators. *Applied Mathematics and Computation* 111 (1): 33–51.

10 Debnath, L. (2012). *Nonlinear Partial Differential Equations for Scientists and Engineers*. New York: Springer Science & Business Media.

18

Preliminaries of Fractal Differential Equations

Fractal analysis is a subject of great interest in various science and engineering applications, especially in computer graphics, architecture [1], medicine [2], fractal antennas [3], etc. Differential equations over fractal domain are often referred to as fractal differential equations. In this context, a basic idea of fractals and notion of fractal differential equations have been incorporated.

18.1 Introduction to Fractal

Nature often exhibits irregularity, nonlinearity, and complexity. Even, the classical Euclidean geometrical shapes viz. lines, circles, quadrilaterals, spheres, etc. lack to abstractly define the complex structures of nature. For instance, the clouds, coastlines, rocks, trees, etc. may not be completely expressed in terms of the classical geometry. In this regard, a new class of geometry was developed by Benoit B. Mandelbrot, mainly referred to as *fractal geometry* (based on a new family of shapes named as *fractals*).

Benoit B. Mandelbrot [4] introduced the concept of fractals and fractal geometry in an imaginative way. A creative bench mark "The Fractal Geometry of Nature" [5] written by Mandelbrot serves as a standard book for elementary concepts of fractals. In a more conventional way, the definition of fractals given by Mandelbrot [5] is stated as,

> A fractal is by definition a set for which the Hausdorff Besicovitch Dimension strictly exceeds the topological dimension.
>
> —B. B. Mandelbrot

Here, Hausdorff Besicovitch dimensional is referred to as the fractional dimension. In this regard, we illuminate glimpses of fractal dimension and

Advanced Numerical and Semi-Analytical Methods for Differential Equations, First Edition.
Snehashish Chakraverty, Nisha Rani Mahato, Perumandla Karunakar, and Tharasi Dilleswar Rao.
© 2019 John Wiley & Sons, Inc. Published 2019 by John Wiley & Sons, Inc.

its distinction from Euclid dimension. Generally, in the Euclidean space, the dimension is integer whereas in the case of fractals the fractal dimension may not be an integer [5, 6]. For a delineated explanation of fractals, we encourage the readers to refer the interesting book [5]. The book brings a clear picture of fractals viz. *Koch curves, snowflakes, Sierpinski gasket (SG)* and its broad range of advanced ideas viz. fractal singularities of differential equations, fractal lattices, self-mapping fractals, randomness, etc. Other standard books for a detailed study of fractals can be found in Refs. [6–10] and further, fractal applications may be found in Refs. [2, 3, 11–13].

In the next section, we discuss few basic shapes of fractal geometry viz. triadic Koch curve, snowflake, and *SG*.

18.1.1 Triadic Koch Curve

Koch curve is one of the earliest and simplest fractal curve constructed using the Euclidean line segment of unit length as given in Table. 18.1.

The figures given in Table 18.1 have been reproduced using the MATLAB code given by D. Scherzinger in http://m2matlabdb.ma.tum.de/download.jsp? MC_ID=5&SC_ID=13&MP_ID=105. Similarly, a Koch snowflake (Figure 18.1) may be generated using an equilateral triangle instead of line segment considered in the Koch curve.

The snowflake given in Figure 18.1 has been reproduced using the MATLAB code given in https://codereview.stackexchange.com/questions/144700/plotting-the-koch-snowflake.

18.1.2 Sierpinski Gasket

SG is named after Polish mathematician W. Sierpinski. The most commonly known *SG* given in Table 18.2 is formed using equilateral triangle by removal of middle inscribed triangle of $\left(\frac{1}{4}\right)$th area. Sometimes, *SG* is also referred to as limit of pre-gaskets G_m as

$$SG = \lim_{m \to \infty} G_m.$$

Figures of G_m and *SG* have been reconstructed through MATLAB exchange file downloaded from https://in.mathworks.com/matlabcentral/fileexchange/50417-sierpinski-fractal-triangle.

Further, the figures of G_m and *SG* have been incorporated into Table 18.2.

Generally, the fractal analysis is not used as a direct approach for solving differential equations. But, the inverse approach of formulating differential equations from fractals is possible using construction of Laplacian on the *SG* and related fractals. In this regard, we have discussed briefly fractal differential equations over *SG* in the next section.

Table 18.1 Triadic Koch curve for $n = 0, 1, 2, 4,$ and 8 iterations.

Triadic Koch curve	Number of segments	Length of	
		Each segment	Koch curve
Number of iterations: 0	1	1	1
Number of iterations: 1	4	$\frac{1}{3}$	$\frac{4}{3}$
Number of iterations: 2	4^2	$\frac{1}{3^2}$	$\left(\frac{4}{3}\right)^2$
Number of iterations: 4	4^3	$\frac{1}{3^3}$	$\left(\frac{4}{3}\right)^3$
Number of iterations: 8	4^4	$\frac{1}{3^4}$	$\left(\frac{4}{3}\right)^4$

18.2 Fractal Differential Equations

A pioneer work on fractal differential equations on *SG* has been carried out by Dalrymple et al. [14]. In this regard, let us illustrate the work of Dalrymple et al. [14] for heat and wave equations in Sections 18.2.1 and 18.2.2, respectively.

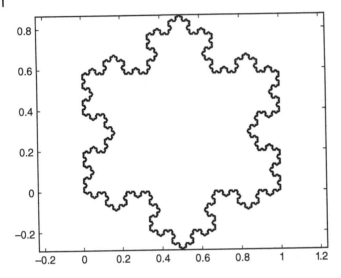

Figure 18.1 Koch snowflake for eight iterations.

18.2.1 Heat Equation

This section discusses the heat equation

$$\frac{\partial u}{\partial t} = \Delta_x u \tag{18.1}$$

subject to Dirichlet boundary conditions

$$u(p_j, t) = 0, \quad j = 0, 1, 2 \tag{18.2}$$

and the initial condition

$$u(x, 0) = f(x) \tag{18.3}$$

where Δ represents the symmetric Laplacian defined by Kigami [15] on a Sierpinski triangle. Here, $u(x, t)$ is a function of $x \in SG$ and $0 \leq t \in \mathfrak{R}$.

The algorithms for computing the approximate solution of Eq. (18.1) may be found in greater detail in Ref. [14]. But, for the sake of completeness, the difference equation obtained using the procedure given in Ref. [14] are reproduced in Eq. (18.4) by restricting time t to integer multiples (in terms of k) for timescale h.

$$u(x, (k + 1)h) - u(x, kh) = h5^m \left(\left(\sum_{(x,y) \in E_m} u(y, kh) \right) - 4u(x, kh) \right) \tag{18.4}$$

where E_m are the edges of pre-gasket G_m.

Table 18.2 Sierpinski triangle for $n = 0, 1, 2, 4$, and 8 iterations.

Number of iterations	Pre-gaskets G_m	Sierpinski triangle
0	G_0	
1	G_1	
2	G_2	
4	G_4	

(Continued)

Table 18.2 (Continued)

Number of iterations	Pre-gaskets G_m	Sierpinski triangle
8	G_8	

18.2.2 Wave Equation

In this section, we consider wave equation

$$\frac{\partial^2 u}{\partial t^2} = \Delta_x u \tag{18.5}$$

subject to boundary conditions $u(p_j, t) = 0; j = 0, 1, 2$ and initial conditions

$$u(x, 0) = f(x) \quad \text{and} \quad \frac{\partial u(x, 0)}{\partial t} = g(x) \tag{18.6}$$

where $x \in SG$ and $0 \leq t \in \Re$. Similar to the heat equation, the difference equation associated with Eq. (18.5) for computation of approximate solution is written as

$$u(x, (k + 2)h) - 2u(x, (k + 1)h) + u(x, kh)$$

$$= h^2 5^m \left(\left(\sum_{(x,y) \in E_m} u(y, kh) \right) - 4u(x, kh) \right) \tag{18.7}$$

using the procedure given in Ref. [14].

It may be noted that the present chapter has been devoted only to illustrate the basic ideas of fractals and differential equations over fractals. As such, Sections 18.2.1 and 18.2.2 just give an overview of application of differential equations over fractals. The detailed analysis of Eqs. (18.1) and (18.5) may be found in Ref. [14]. Further, readers interested in having a thorough knowledge of fractal differential equations and fractal-time dynamical systems are highly encouraged to refer Refs. [16, 17].

References

1 Ostwald, M.J. and Vaughan, J. (2016). *The Fractal Dimension of Architecture*, vol. 1. Cham: Birkhäuser.

2 Losa, G.A. (2011). Fractals in biology and medicine. In: *Reviews in Cell Biology and Molecular Medicine*, vol. 4. (ed. R.A. Meyers), Wiley Online Library. doi: 10.1002/3527600906.mcb.201100002.

3 Hohlfeld, R.G. and Cohen, N. (1999). Self-similarity and the geometric requirements for frequency independence in antennae. *Fractals* 7 (1): 79–84.

4 Mandelbrot, B. (1977). *Fractals*, 24. San Francisco: W.H. Freeman.

5 Mandelbrot, B.B. (1982). *The Fractal Geometry of Nature*, vol. 1. New York: W.H. Freeman.

6 Feder, J. (2013). *Fractals*. New York: Springer Science & Business Media.

7 Falconer, K. (2004). *Fractal Geometry: Mathematical Foundations and Applications*. Hoboken, NJ: Wiley.

8 Barnsley, M.F., Devaney, R.L., Mandelbrot, B.B. et al. (2012). In: *The Science of Fractal Images* (ed. H.O. Peitgen and D. Saupe). New York: Springer-Verlag.

9 Barnsley, M.F. (2014). *Fractals Everywhere*. Boston: Academic Press.

10 Peitgen, H.O., Jürgens, H., and Saupe, D. (2006). *Chaos and Fractals: New Frontiers of Science*. New York: Springer Science & Business Media.

11 Dimri, V.P. (ed.) (2000). *Application of Fractals in Earth Sciences*. Rotterdam: A. A. Balkema.

12 Heck, A. and Perdang, J.M. (eds.) (2008). *Applying Fractals in Astronomy*, vol. 3. Berlin, Heidelberg: Springer-Verlag.

13 Ghanbarian, B. and Hunt, A.G. (eds.) (2017). *Fractals: Concepts and Applications in Geosciences*. Boca Raton, FL: CRC Press.

14 Dalrymple, K., Strichartz, R.S., and Vinson, J.P. (1999). Fractal differential equations on the Sierpinski gasket. *Journal of Fourier Analysis and Applications* 5 (2–3): 203–284.

15 Kigami, J. (1989). A harmonic calculus on the Sierpinski spaces. *Japan Journal of Applied Mathematics* 6 (2): 259–290.

16 Parvate, A. and Gangal, A.D. (2005). Fractal differential equations and fractal-time dynamical systems. *Pramana* 64 (3): 389–409.

17 Strichartz, R.S. (2006). *Differential Equations on Fractals: A Tutorial*. Princeton, NJ: Princeton University Press.

19

Differential Equations with Interval Uncertainty

19.1 Introduction

Generally, differential equations are the backbone of various physical systems occurring in a wide variety of disciplines viz. physics, chemistry, biology, economics, and engineering [1, 2]. These physical systems are modeled either by ordinary or partial differential equations. Generally, in differential equations, the involved coefficients and variables are considered as deterministic or exact values. In that case, one may handle such differential equations (with deterministic coefficients or variables) by known analytic or numerical methods [3–5].

In actual practice, due to errors in experimental observation or due to truncation of the parametric values, etc. we may have only imprecise, insufficient, or incomplete information about the involved parameters of the differential equations. So, the parametric values involved in such differential equations are uncertain in nature. As such, there is a need of modeling different physical problems with uncertain parameters. In general, these uncertainties may be modeled through probabilistic, interval, or fuzzy approach [4, 5]. Probabilistic methods may not be able to deliver reliable results at the required condition without sufficient data. Therefore, in the recent years, interval analysis and fuzzy set theory have become powerful tools for uncertainty modeling.

In this chapter, we present different approaches to handle differential equations with interval uncertainty.

19.2 Interval Differential Equations

Moore [6] first introduced the concept of interval analysis and computations in 1966. Thereafter, this concept has been successfully applied by various researchers for uncertainty analysis. Several books [6–8] have also been written by different authors representing the scope and various aspects of interval analysis. These books give an extensive review of interval computations which may help the readers to understand the basic concepts of interval analysis.

Advanced Numerical and Semi-Analytical Methods for Differential Equations, First Edition.
Snehashish Chakraverty, Nisha Rani Mahato, Perumandla Karunakar, and Tharasi Dilleswar Rao.
© 2019 John Wiley & Sons, Inc. Published 2019 by John Wiley & Sons, Inc.

Various mathematical techniques have been developed by different authors to handle the differential equations with interval analysis. As such, differential calculus is studied by Chalco-Cano et al. [9] for interval-valued functions using generalized Hukuhara differentiability [10], which is often referred as the most general concept of differentiability for interval-valued functions. Also, the Hukuhara concept has been utilized by Stefanini and Bede [10, 11] in a more generalized way for interval-valued functions and interval differential equations (IDEs). Differential transformation method (DTM) is applied by Ghazanfaria and Ebrahimia [12]. Recently, Chakraverty et al. developed new techniques and approaches to handle differential equations with uncertain parameters [13–15].

The next section presents the preliminary concepts of interval arithmetic and parametric form of interval numbers.

19.2.1 Interval Arithmetic

An interval is denoted as $\widetilde{X} = [\underline{X}, \ \overline{X}]$, where $\underline{X}$ and $\overline{X}$ represent the lower and upper bounds of interval $\widetilde{X}$, respectively. Any two intervals $\widetilde{X}$ and $\widetilde{Y}$ are considered to be equal if their corresponding bounds are equal. The basic interval arithmetic operations are as follows [6, 7]:

1. $\widetilde{X} + \widetilde{Y} = [\underline{X} + \underline{Y}, \overline{X} + \overline{Y}]$
2. $\widetilde{X} - \widetilde{Y} = [\underline{X} - \overline{Y}, \overline{X} - \underline{Y}]$
3. $\widetilde{X} \times \widetilde{Y} = [\min(\underline{X} \times \overline{Y}, \underline{X} \times \underline{Y}, \overline{X} \times \underline{Y}, \overline{X} \times \overline{Y}), \max(\underline{X} \times \overline{Y}, \underline{X} \times \underline{Y}, \overline{X} \times \underline{Y}, \overline{X} \times \overline{Y})]$
4. $\frac{\widetilde{X}}{\widetilde{Y}} = \left[\min\left(\frac{\underline{X}}{\underline{Y}}, \frac{\underline{X}}{\overline{Y}}, \frac{\overline{X}}{\underline{Y}}, \frac{\overline{X}}{\overline{Y}} \right), \max\left(\frac{\underline{X}}{\underline{Y}}, \frac{\underline{X}}{\overline{Y}}, \frac{\overline{X}}{\underline{Y}}, \frac{\overline{X}}{\overline{Y}} \right) \right], \underline{Y}, \overline{Y} \neq 0$
5. $X_\nabla = \frac{(\overline{X} - \underline{X})}{2}$ (Radius of interval $\widetilde{X}$)
6. $X_c = \frac{(\overline{X} + \underline{X})}{2}$ (Centre of interval $\widetilde{X}$)
7. $k\widetilde{X} = \begin{cases} [k\underline{X}, k\overline{X}], & k \geq 0 \\ [k\overline{X}, k\underline{X}], & k < 0 \end{cases}$ where $k \in \Re$

19.3 Generalized Hukuhara Differentiability of IDEs

One of the known definitions of difference and derivatives for interval-valued functions was given by Hukuhara [16]. Further, the new concepts named as generalized Hukuhara differences have been examined in Refs. [11, 12].

Let us consider two intervals $\widetilde{A} = [\underline{a}, \overline{a}]$ and $\widetilde{B} = [\underline{b}, \overline{b}]$. Then, the Hukuhara difference [11] is defined as

$$\widetilde{A} - \widetilde{B} = \widetilde{C} \Rightarrow [\underline{a}, \overline{a}] - [\underline{b}, \overline{b}] = [\underline{c}, \overline{c}]$$

where $\underline{c} = \min\{\underline{a} - \underline{b}, \overline{a} - \overline{b}\}$ and $\overline{c} = \max\{\underline{a} - \underline{b}, \overline{a} - \overline{b}\}$.

19.3.1 Modeling IDEs by Hukuhara Differentiability

Let us consider a linear interval-valued differential equation [11, 12] as

$$\tilde{y}' = \tilde{f}(x, y), \quad \tilde{y}(x_0) = \tilde{y}_0 \tag{19.1}$$

where

$$\tilde{f}(x, y) = \left[\underline{f}(x, y), \overline{f}(x, y)\right] \quad \text{for} \quad \tilde{y} = \left[\underline{y}, \overline{y}\right], \tilde{y}_0 = \left[\underline{y}_0, \overline{y}_0\right].$$

19.3.1.1 Solving by Integral Form

After applying integration to Eq. (19.1) the equivalent integral equation [10, 11] is obtained as

$$\tilde{y}(x) - \tilde{y}_0 = \int_{x_0}^{x_0 + \delta} \tilde{f}(t, y(t)) dt \tag{19.2}$$

where δ is a small positive quantity.

Using the Hukuhara difference

$$\tilde{y}(x) - \tilde{y}_0 = \left[\min\left\{\underline{y}(x) - \underline{y}_0, \overline{y}(x) - \overline{y}_0\right\}, \max\left\{\underline{y}(x) - \underline{y}_0, \overline{y}(x) - \overline{y}_0\right\}\right]$$

Now, according to Eq. (19.2), one may obtain

$$\left. \begin{array}{l} \min\left\{\underline{y}(x) - \underline{y}_0, \overline{y}(x) - \overline{y}_0\right\} = \int_{x_0}^{x_0 + \delta} \underline{f}(t, y(t)) dt, \\[2mm] \max\left\{\underline{y}(x) - \underline{y}_0, \overline{y}(x) - \overline{y}_0\right\} = \int_{x_0}^{x_0 + \delta} \overline{f}(t, y(t)) dt. \end{array} \right\} \tag{19.3}$$

19.3.1.2 Solving by Differential Form

By using the differential form [12, 13] to Eq. (19.1), we obtain

$$\left. \begin{array}{l} \min\left\{\underline{y}'(x), \overline{y}'(x)\right\} = \underline{f}(x, y) \\[2mm] \max\left\{\underline{y}'(x), \overline{y}'(x)\right\} = \overline{f}(x, y) \end{array} \right\} \tag{19.4}$$

with respect to the initial conditions $y(x_0) = \underline{y}_0, \overline{y}(x_0) = \overline{y}_0$.

We obtain two situations from Eq. (19.4) as

Case (i): If $\underline{y}'(x) \leq \overline{y}'(x)$, then the possible differential equations result in

$$\left. \begin{array}{l} \underline{y}'(x) = \underline{f}(x, y), \\[2mm] \overline{y}'(x) = \overline{f}(x, y). \end{array} \right\} \tag{19.5}$$

with respect to the initial conditions, $y(x_0) = \underline{y}_0, \overline{y}(x_0) = \overline{y}_0$.

Case (ii): If $y'(x) \geq \bar{y}'(x)$, then the differential equations reduce to

$$\left.\begin{array}{l} \underline{y}'(x) = \bar{f}(x, y), \\ \bar{y}'(x) = \underline{f}(x, y). \end{array}\right\} \tag{19.6}$$

with respect to the initial conditions, $\underline{y}(x_0) = \underline{y}_0$, $\bar{y}(x_0) = \bar{y}_0$.

Example 19.1 Let us consider an IDE

$$\begin{cases} \tilde{y}' = -2\tilde{y} + [1,3]x \\ \tilde{y}(0) = [1,2] \end{cases} \tag{19.7}$$

Solution By using the Hukuhara differentiation approach given in Eqs. (19.5) and (19.6), one may obtain the following two cases as explained above:
Case (i):

$$\left.\begin{array}{l} \underline{y}' = -2\bar{y} + x \\ \bar{y}' = -2y + 3x \\ \underline{y}(0) = 1, \quad \bar{y}(0) = 2 \end{array}\right\} \tag{19.8}$$

and
Case (ii):

$$\left.\begin{array}{l} \underline{y}' = -2y + 3x \\ \bar{y}' = -2\bar{y} + x \\ \underline{y}(0) = 1, \quad \bar{y}(0) = 2 \end{array}\right\} \tag{19.9}$$

As such, one may conclude that Eq. (19.7) has exactly two solutions. One of them starts with the case (i) of differentiability

$$\left[\underline{y}(x), \bar{y}(x)\right] = \left[\frac{-3}{4}e^{2x} + 2e^{-2x} + \frac{3x}{2} - \frac{1}{4}, \frac{3}{4}e^{2x} + 2e^{-2x} + \frac{x}{2} - \frac{3}{4}\right]$$

The other starts with case (ii) as

$$\left[\underline{y}(x), \bar{y}(x)\right] = \left[-\frac{3x}{2} + \frac{7}{4}e^{2x} - \frac{3}{4}, -\frac{x}{2} + \frac{9}{4}e^{2x} - \frac{1}{4}\right]$$

Accordingly, the interval bounds of $\tilde{y}(x)$ is compared with trivial solutions in Figure 19.1.

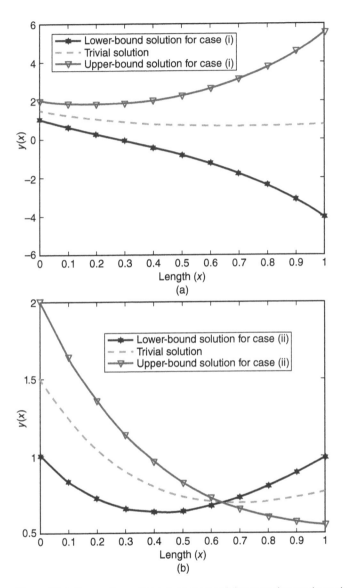

Figure 19.1 Comparison of trivial solution with lower and upper bounds obtained using (a) case (i) and (b) case (ii).

19.4 Analytical Methods for IDEs

As mentioned in the introduction, several numerical and analytical methods [14, 17] are available to solve IDEs. Generally, analytical solutions have a significant role in proper understanding of various science and engineering problems [18]. Accordingly, below we present one of the new approaches to handle IDEs [14, 15, 17].

19.4.1 General form of *n*th-order IDEs

Let us consider *n*th-order IDE in general form given in Refs. [14, 17] as

$$\widetilde{y}^{n}(x) + c_{n-1}\widetilde{y}^{n-1}(x) + \cdots + c_{1}\widetilde{y}^{1}(x) + c_{0}\widetilde{y}(x) = \widetilde{R}(x) \tag{19.10}$$

subject to the initial conditions

$$\widetilde{y}(0) = \widetilde{b}_{0}, \widetilde{y}^{1}(0) = \widetilde{b}_{1}, \ldots, \widetilde{y}^{n-1}(0) = \widetilde{b}_{n-1}$$

where c_i's are real constants and $\widetilde{b}_i$'s are interval values for $0 \leq i \leq n-1$.
Here, $\widetilde{y}(x)$ is the solution to be determined.

Now, Eq. (19.10) may be represented in terms of lower and upper bounds as

$$\left[\underline{y}^{n}(x), \overline{y}^{n}(x)\right] + c_{n-1}\left[\underline{y}^{n-1}(x), \overline{y}^{n-1}(x)\right] + \cdots + c_{1}\left[\underline{y}'(x), \overline{y}'(x)\right]$$
$$+ c_{0}\left[\underline{y}(x), \overline{y}(x)\right] = \left[\underline{R}(x), \overline{R}(x)\right] \tag{19.11}$$

subject to interval initial conditions

$$[\underline{y}(0), \overline{y}(0)] = \left[\underline{b}_{0}, \overline{b}_{0}\right], \left[\underline{y}'(0), \overline{y}'(0)\right] = \left[\underline{b}_{1}, \overline{b}_{1}\right], \ldots, \left[\underline{y}^{(n-1)}(0), \overline{y}^{(n-1)}(0)\right]$$
$$= \left[\underline{b}_{n-1}, \overline{b}_{n-1}\right]$$

From Eq. (19.11), three possible cases arise with respect to the sign of the involved coefficients. The possible three cases are given as [17, 18]:

Case (i): All the coefficients $c_{n-1}, c_{n-2}, \ldots, c_1, c_0$ are positive.
Case (ii): Coefficients $c_{n-1} \cdots c_{n-m}$ are positive and $c_{n-m-1}, c_{n-m-2}, \ldots, c_1, c_0$ are negative.
Case (iii): All coefficients $c_{n-1}, c_{n-2}, \ldots, c_1, c_0$ are negative.

The analytical methods to handle IDE (19.10) in all three possible cases mentioned above are presented in the next section.

19.4.2 Method Based on Addition and Subtraction of Intervals

In this section, we present the solution procedure based on addition and subtraction of intervals [18] to handle all the three possible cases of considered IDE given by Eq. (19.10).

Case (i): When all the coefficients $c_{n-1}, c_{n-2}, \ldots, c_1, c_0$ involved in Eq. (19.10) are positive. Then, one may obtain the following equations in terms of lower and upper bound as

$$\underline{y}^{n}(x) + c_{n-1}\underline{y}^{n-1}(x) + \cdots + c_{1}\underline{y}'(x) + c_{0}\underline{y}(x) = \underline{R}(x), \tag{19.12}$$

and

$$\overline{y}^n(x) + c_{n-1}\overline{y}^{n-1}(x) + \cdots + c_1\overline{y}'(x) + c_0\overline{y}(x) = \overline{R}(x). \tag{19.13}$$

By adding Eqs. (19.12) and (19.13), one may obtain

$$\left(\underline{y}^n(x) + \overline{y}^n(x)\right) + c_{n-1}\left(\underline{y}^{n-1}(x) + \overline{y}^{n-1}(x)\right) + \cdots + c_1\left(\underline{y}'(x) + \overline{y}'(x)\right)$$
$$+ c_0\left(\underline{y}(x) + \overline{y}(x)\right) = \left(\underline{R}(x) + \overline{R}(x)\right) \tag{19.14}$$

subject to the initial conditions

$$\left(\underline{y} + \overline{y}\right)(0) = \underline{b}_0 + \overline{b}_0, \left(\underline{y}' + \overline{y}'\right)(0) = \underline{b}_1 + \overline{b}_1, \ldots, \left(\underline{y}^{(n-1)} + \overline{y}^{(n-1)}\right)(0)$$
$$= \underline{b}_{n-1} + \overline{b}_{n-1}.$$

For simplification, Eq. (19.14) is represented as

$$u^n(x) + c_{n-1}u^{n-1}(x) + \cdots + c_1(t)u'(x) + c_0u(x) = f(x), \tag{19.15}$$

with respect to the initial conditions

$$u(0) = l_0, \quad u'(0) = l_1, \ldots, u^{n-1}(0) = l_{n-1}$$

where $\underline{y}^i(x) + \overline{y}^i(x) = u^i(x), 0 \le i \le n$, $\underline{R}(x) + \overline{R}(x) = f(x)$, $\underline{y}^i(0) + \overline{y}^i(0) = u^i(0)$, and $\underline{b}_i + \overline{b}_i = l_i, 0 \le i \le n-1$.

Similarly, subtracting Eqs. (19.12) and (19.13), one may obtain

$$\left(\underline{y}^n(x) - \overline{y}^n(x)\right) + c_{n-1}\left(\underline{y}^{n-1}(x) - \overline{y}^{n-1}(x)\right) + \cdots + c_1\left(\underline{y}'(x) - \overline{y}'(x)\right)$$
$$+ c_0\left(\underline{y}(x) - \overline{y}(x)\right) = \left(\underline{R}(x) - \overline{R}(x)\right) \tag{19.16}$$

Further, Eq. (19.16) is represented as

$$v^n(x) + c_{n-1}v^{n-1}(x) + \cdots + c_1(t)v'(x) + c_0v(x) = g(x) \tag{19.17}$$

with respect to the initial conditions

$$v(0) = m_0, \quad v'(0) = m_1, \ldots, v^{n-1}(0) = m_{n-1}$$

where $\underline{y}^i(x) - \overline{y}^i(x) = v^i(x)$, $0 \le i \le n$, $\underline{R}(x) - \overline{R}(x) = g(x)$, and $\underline{y}^i(0) - \overline{y}^i(0) = v^i(0)$,

$$\underline{b}_i - \overline{b}_i = m_i, \quad 0 \le i \le n-1.$$

It may be noted that both Eqs. (19.15) and (19.17) are in crisp forms. So, one may easily obtain solutions

$$u(x) = \underline{y}(x) + \overline{y}(x) \quad \text{and} \quad v(x) = \underline{y}(x) - \overline{y}(x).$$

Then, by applying standard methods, the lower-bound solution is obtained by adding $u(x)$ and $v(x)$ whereas the upper-bound solution is obtained by subtracting $u(x)$ and $v(x)$.

Case (ii): When some of the coefficients $c_{n-1},\ldots, c_{n-m}$ are positive and $c_{n-m-1}, c_{n-m-2},\ldots, c_1, c_0$ are negative [14, 18] in Eq. (19.10), one may obtain the lower- and upper-bound equations as

$$\underline{y}''(x) + c_{n-1}\underline{y}^{n-1}(x) + \cdots + c_{n-m}\underline{y}^{n-m}(x) + c_{n-m-1}\overline{y}^{n-m-1}(x) + \cdots + c_0\overline{y}(x) = \underline{R}(x),$$

(19.18)

and

$$\overline{y}''(x) + c_{n-1}\overline{y}^{n-1}(x) + \cdots + c_{n-m}\overline{y}^{n-m}(x) + c_{n-m-1}\underline{y}^{n-m-1}(x) + \cdots + c_0\underline{y}(x) = \overline{R}(x)$$

(19.19)

subject to the initial conditions

$$\left[\underline{y}(0),\overline{y}(0)\right] = \left[\underline{b}_0,\overline{b}_0\right], \left[\underline{y}'(0),\overline{y}'(0)\right] = \left[\underline{b}_1,\overline{b}_1\right], \ldots, \left[\underline{y}^{(n-1)}(0),\overline{y}^{(n-1)}(0)\right]$$
$$= \left[\underline{b}_{n-1},\overline{b}_{n-1}\right]$$

Now, by using the same procedure as in case (i) one may also obtain the required lower- and upper-bound solutions for this case.

Case (iii): When all the coefficients $c_{n-1}, c_{n-2},\ldots, c_1, c_0$ involved in Eq. (19.10) are negative. Then, one may obtain the following lower- and upper-bound equations [18, 19] as

$$\underline{y}''(x) + c_{n-1}\overline{y}^{n-1}(x) + \cdots + c_1\overline{y}'(x) + c_0\overline{y}(x) = \underline{R}(x)$$

(19.20)

and

$$\overline{y}''(x) + c_{n-1}\underline{y}^{n-1}(x) + \cdots + c_1\underline{y}'(x) + c_0\underline{y}(x) = \overline{R}(x)$$

(19.21)

subject to the initial conditions

$$\left[\underline{y}(0),\overline{y}(0)\right] = \left[\underline{b}_0,\overline{b}_0\right], \left[\underline{y}'(0),\overline{y}'(0)\right] = \left[\underline{b}_1,\overline{b}_1\right], \ldots, \left[\underline{y}^{(n-1)}(0),\overline{y}^{(n-1)}(0)\right]$$
$$= \left[\underline{b}_{n-1},\overline{b}_{n-1}\right]$$

Similarly, as in case (i), one may obtain the required lower- and upper-bound solutions for case (iii). For more detailed understanding of various analytical methods to handle IDEs, one may refer Refs. [14, 15, 17, 19, 20].

Example 19.2 Find the lower- and upper-bound solutions [14, 18] of the IDE $\widetilde{y}'' - 5\widetilde{y}' + 4\widetilde{y} = 0$, subject to the initial conditions $\widetilde{y}(0) = [0.2, 0.8]$ and $\widetilde{y}'(0) = [0.8, 1.2]$.

Solution The given equation may be expressed as

$$\left[\underline{y''},\overline{y''}\right] - 5\left[\underline{y'},\overline{y'}\right] + 4\left[\underline{y},\overline{y}\right] = 0. \tag{19.22}$$

From Eq. (19.22), the lower- and upper-bound equations are obtained as

$$\underline{y''} - 5\overline{y'} + 4\underline{y} = 0 \tag{19.23}$$

and

$$\overline{y''} - 5\underline{y'} + 4\overline{y} = 0. \tag{19.24}$$

Now, by adding Eqs. (19.23) and (19.24), we obtain

$$u'' - 5u' + 4u = 0, \tag{19.25}$$

subject to the initial conditions

$$u(0) = 1, \quad u'(0) = 2,$$

where $u(x) = \underline{y}(x) + \overline{y}(x)$.

The general solution of Eq. (19.25) is obtained as

$$u(x) = \frac{2}{3}e^x + \frac{1}{3}e^{4x} \tag{19.26}$$

Similarly, by subtracting Eqs. (19.23) and (19.24), we get

$$v'' + 5v' + 4v = 0, \tag{19.27}$$

subject to the initial conditions

$$v(0) = -0.6 \quad \text{and} \quad v'(0) = -0.4.$$

where $v(x) = \underline{y}(x) - \overline{y}(x)$.

The general solution of Eq. (19.27) is further obtained as

$$v(x) = -\frac{2.8}{3}e^{-x} + \frac{1}{3}e^{-4x}. \tag{19.28}$$

From Eqs. (19.26) and (19.28), the lower- and upper-bound solutions of Eq. (19.22) may be obtained as

$$\underline{y}(x) = \frac{1}{3}(2\sinh(x) + \cosh(4x)) - \frac{2}{15}e^{-x} \quad \text{and}$$

$$\overline{y}(x) = \frac{1}{3}(2\cosh(x) + \sinh(4x)) + \frac{2}{15}e^{-x}.$$

The solution bounds are depicted in Figure 19.2.

Detailed explanations regarding IDEs may be found in Refs. [10–20] and the references mentioned therein.

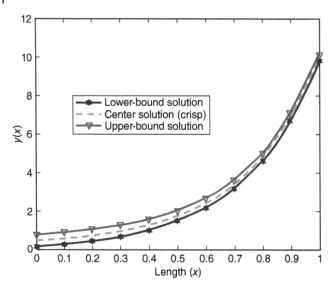

Figure 19.2 Lower-, center-, and upper-bound solutions of Example 19.2.

Exercise

1 By using Hukuhara differentiation, solve the following problems:

(i)
$$\begin{cases} \widetilde{y}' = -3\widetilde{y} + [5, 10]x \\ \widetilde{y}(0) = [7, 11] \end{cases}$$

(ii)
$$\begin{cases} \widetilde{y}' + \left(\dfrac{1}{x}\right)\widetilde{y} = [1, 4]x^2 \\ \widetilde{y}(0) = [0, 1] \end{cases}$$

2 Solve the following IDEs by using addition and subtraction of interval numbers
(i) $\widetilde{y}''' - 2\widetilde{y}'' + 4\widetilde{y}' + \widetilde{y} = x$ with respect to $\widetilde{y}(0) = [0.1, 0.7]$, $\widetilde{y}'(0) = [0.2, 1.1]$, and $\widetilde{y}''(0) = [0.2, 1.1]$
(ii) $\widetilde{y}'' - \widetilde{y}' + 4\widetilde{y} = x \sin x$ with respect to $\widetilde{y}(0) = [0.2, 0.4], \widetilde{y}'(0) = [0.3, 2.1]$

References

1 Hale, J.K. and Lunel, S.M.V. (2013). *Introduction to Functional Differential Equations*, vol. 99. New York: Springer Science & Business Media.

2 Farkas, M. (1977). *Differential Equations*. Amsterdam: North-Holland Publishing Co.

3 Bhat, R.B. and Chakraverty, S. (2004). *Numerical Analysis in Engineering*. Oxford: Alpha Science International.

4 Chakraverty, S., Sahoo, B.K., Rao, T.D. et al. (2018). Modelling uncertainties in the diffusion-advection equation for radon transport in soil using interval arithmetic. *Journal of Environmental Radioactivity* 182: 165–171.

5 Rao, T.D. and Chakraverty, S. (2017). Modeling radon diffusion equation in soil pore matrix by using uncertainty based orthogonal polynomials in Galerkin's method. *Coupled Systems Mechanics* 6 (4): 487–499.

6 Moore, R.E. (1966). *Interval analysis*, Series in Automatic Computation. Englewood Cliff, NJ: Prentice-Hall.

7 Moore, R.E., Kearfott, R.B., and Cloud, M.J. (2009). *Introduction to Interval Analysis*, vol. 110. Philadelphia, PA: SIAM.

8 Alefeld, G. and Herzberger, J. (2012). *Introduction to Interval Computation*. New York: Academic Press.

9 Jaulin, L., Kieffer, M., Didrit, O., and Walter, E. (2001). *Applied Interval Analysis: With Examples in Parameter and State Estimation, Robust Control and Robotics*, vol. 1. London: Springer Science & Business Media.

10 Chalco-Cano, Y., Rufián-Lizana, A., Román-Flores, H., and Jiménez-Gamero, M.D. (2013). Calculus for interval-valued functions using generalized Hukuhara derivative and applications. *Fuzzy Sets and Systems* 219: 49–67.

11 Stefanini, L. and Bede, B. (2009). Generalized Hukuhara differentiability of interval-valued functions and interval differential equations. *Nonlinear Analysis: Theory, Methods & Applications* 71 (3–4): 1311–1328.

12 Stefanini, L. and Bede, B. (2012). Some notes on generalized Hukuhara differentiability of interval-valued functions and interval differential equations (No. 1208). University of Urbino Carlo Bo, Italy Department of Mathematics, WP-EMS.

13 Ghazanfaria, B. and Ebrahimia, P. (2013). Differential transformation method for solving interval differential equations. *Journal of Novel Applied Sciences* 2: 598–604.

14 Tapaswini, S. and Chakraverty, S. (2014). New analytical method for solving n-th order fuzzy differential equations. *Annals of Fuzzy Mathematics and Informatics* 8: 231–244.

15 Chakraverty, S. (ed.) (2014). *Mathematics of Uncertainty Modeling in the Analysis of Engineering and Science Problems*. Hershey, PA: IGI Global.

16 Hukuhara, M. (1967). Integration des applications mesurables dont la valeur est un compact convexe. *Funkcialaj Ekvacioj* 10 (3): 205–223.

17 Tapaswini, S., Chakraverty, S., and Allahviranloo, T. (2017). A new approach to nth order fuzzy differential equations. *Computational Mathematics and Modeling* 28 (2): 278–300.

18 Chakraverty, S., Tapaswini, S., and Behera, D. (2016). *Fuzzy Differential Equations and Applications for Engineers and Scientists.* Boca Raton, FL: CRC Press.

19 Corliss, G.F. (1995). Guaranteed error bounds for ordinary differential equations. In: *Theory of Numerics in Ordinary and Partial Differential Equations* (ed. M. Ainsworth, J. Levesley, W.A. Light and M. Marletta). Oxford: Oxford University Press.

20 Rihm, R. (1994). Interval methods for initial value problems in ODEs. In: *Topics in Validated Computations* (ed. J. Herzberger), 173–207. Amsterdam: Elsevier.

20

Differential Equations with Fuzzy Uncertainty

20.1 Introduction

In this chapter, a system of fuzzy linear differential equations is studied. Recently, a new technique using the triangular fuzzy numbers (TFNs) [1, 2] is illustrated to model the fuzzy linear differential equations. The solution of linear differential equations with fuzzy initial conditions may be studied as a set of intervals by varying α-cut. The term fuzzy differential equations were first introduced by Chang and Zadeh [3, 4]. Later, Bede introduced a strongly generalized differentiability of fuzzy functions in Refs. [5, 6]. Recently, various researchers viz. Allahviranloo et al. [7], Chakraverty et al. [8], Tapaswini and Chakraverty [9] have studied fuzzy differential equations. As such, a geometric approach to solve fuzzy linear systems of differential equations have been studied by Gasilov et al. [1, 2, 10, 11]. The difference between this method and the methods offered to handle the system of fuzzy linear differential equation is that at any time the solution consists a fuzzy region in the coordinate space. In this regard, the following section presents a procedure to solve fuzzy linear system of differential equations.

20.2 Solving Fuzzy Linear System of Differential Equations

20.2.1 α-Cut of TFN

To understand the preliminary concepts of fuzzy set theory, one can refer Refs. [12, 13]. There exist various types of fuzzy numbers and among them the TFN is found to be mostly used by different authors.

A TFN $\widetilde{X} = (a, c, b)$ is a convex normalized fuzzy set $\widetilde{X}$ of the real line $\Re$ [3, 4] such that,

Advanced Numerical and Semi-Analytical Methods for Differential Equations, First Edition.
Snehashish Chakraverty, Nisha Rani Mahato, Perumandla Karunakar, and Tharasi Dilleswar Rao.
© 2019 John Wiley & Sons, Inc. Published 2019 by John Wiley & Sons, Inc.

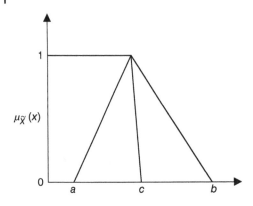

Figure 20.1 Triangular fuzzy number.

1. $\exists$ exactly one $x_0 \in R$ with $\mu_{\widetilde{X}}(x_0) = 1$ where x_0 is referred to as the mean value of $\widetilde{X}$ and $\mu_{\widetilde{X}}$ is the membership function of the fuzzy set such that $\mu_{\widetilde{X}} \in [0, 1]$.
2. $\mu_{\widetilde{X}}$ is piecewise continuous.

The membership function $\mu_{\widetilde{X}}$ of $\widetilde{X}$ is defined (Figure 20.1) as follows:

$$\mu_{\widetilde{X}}(x) = \begin{cases} 0, & x \le a \\ \dfrac{x-a}{c-a}, & a \le x \le c \\ \dfrac{x-b}{c-b}, & c \le x \le b \\ 0, & x \ge b \end{cases}$$

where $c \ne a$ and $c \ne b$.

Using the α-cut approach, the TFN $\widetilde{X} = (a, c, b)$ can further be represented by an ordered pair of functions, $[\underline{x}(\alpha), \overline{x}(\alpha)] = [a + (c - a)\alpha, b + (c - b)\alpha]$, where $\alpha \in [0, 1]$.

20.2.2 Fuzzy Linear System of Differential Equations (FLSDEs)

Let a_{ij} are crisp numbers, $h_i(t)$ are given crisp functions, $\widetilde{b}_i = [\underline{b}_i(\alpha), \overline{b}_i(\alpha)]$ for $0 \le \alpha \le 1$, and $1 \le i, j \le n$, (are fuzzy numbers). Fuzzy Linear System of Differential Equations (FLSDEs) [1, 2] may be written as

$$\left.\begin{aligned} x_1'(t) &= a_{11}x_1 + a_{12}x_2 + \cdots + a_{1n}x_n + h_1(t) \\ x_2'(t) &= a_{21}x_1 + a_{22}x_2 + \cdots + a_{2n}x_n + h_2(t) \\ &\vdots \\ x_n'(t) &= a_{n1}x_1 + a_{n2}x_2 + \cdots + a_{nn}x_n + h_n(t) \end{aligned}\right\} \tag{20.1}$$

subject to the fuzzy initial conditions

$$x_1(t_0) = \widetilde{b}_1, x_2(t_0) = \widetilde{b}_2, \ldots, x_n(t_0) = \widetilde{b}_n. \tag{20.2}$$

One can formulate Eqs. (20.1) and (20.2) in matrix notation as follows:

$$\left.\begin{array}{l} X' = AX + H(t) \\ X(t_0) = \widetilde{B} \end{array}\right\} \tag{20.3}$$

where $A = (a_{ij})$ is $n \times n$ crisp matrix, $H(t) = (h_1(t), h_2(t), \ldots, h_n(t))^T$ is the crisp vector function, and $\widetilde{B} = (\widetilde{b}_1, \widetilde{b}_2, \ldots, \widetilde{b}_n)^T$ is the vector of fuzzy numbers.

20.2.3 Solution Procedure for FLSDE

Let us express the initial condition vector as $\widetilde{B} = (b_{cr})_i + (\widetilde{b})_i$, where $(b_{cr})_i$ denotes the crisp part of $\widetilde{B}$ and $(\widetilde{b})_i$ denotes the uncertain part. Solution of such system may be considered of the form $\widetilde{X}(t) = x_{cr}(t) + \widetilde{x}(t)$ (crisp solution + uncertainty). Here, $x_{cr}(t)$ is a solution of the nonhomogeneous crisp problem as given in Refs. [1, 2]

$$\left.\begin{array}{l} X' = AX + H(t) \\ X(t_0) = (b_{cr})_i \end{array}\right\} \tag{20.4}$$

whereas $\widetilde{x}(t)$ is the solution of the homogeneous system with fuzzy initial conditions,

$$\left.\begin{array}{l} X' = AX \\ X(t_0) = (\widetilde{b})_i. \end{array}\right\} \tag{20.5}$$

It is possible to compute $x_{cr}(t)$ by means of known analytical or numerical methods. Our aim is to solve Eq. (20.3) with fuzzy initial conditions. In this regard, fuzzy uncertainty in terms of TFN is taken into consideration. Let $\widetilde{B} = (p_i, q_i, r_i)$ be the TFN, then $\widetilde{B}$ is written as $(b_{cr})_i + (\widetilde{b})_i = q_i + (\underline{b}_i, 0, \overline{b}_i)$, where q_i designates the crisp part and $(\underline{b}_i, 0, \overline{b}_i)$ designates the TFN.

Below, we present an example problem for clear understanding of the described method.

Example 20.1   Solve the system $\begin{bmatrix} x' \\ y' \end{bmatrix} = \begin{bmatrix} \frac{6}{7} & -\frac{15}{14} \\ 5 & \frac{37}{14} \end{bmatrix} \begin{bmatrix} x \\ y \end{bmatrix} + \begin{bmatrix} e^{2t} \\ e^{-t} \end{bmatrix}$ with the initial

TFN values $\begin{bmatrix} x(0) \\ y(0) \end{bmatrix} = \begin{bmatrix} (13.5, 14, 15) \\ (3, 7, 10) \end{bmatrix}$

Solution The system is in the form as given in Eq. (20.3),

$$
\left.
\begin{array}{l}
X' = AX + H(t) \\
X(0) = \widetilde{B}
\end{array}
\right\}
$$

where $X' = \begin{bmatrix} x' \\ y' \end{bmatrix}$, $A = \begin{bmatrix} \frac{6}{7} & -\frac{15}{14} \\ 5 & \frac{37}{14} \end{bmatrix}$, $X = \begin{bmatrix} x \\ y \end{bmatrix}$, $H(t) = \begin{bmatrix} e^{2t} \\ e^{-t} \end{bmatrix}$ and $X(0) = \begin{bmatrix} x(0) \\ y(0) \end{bmatrix}$.

Here, the initial conditions $\widetilde{B}$ are TFNs,

$$
\widetilde{B} = \begin{bmatrix} (13.5, 14, 15) \\ (3, 7, 10) \end{bmatrix} = \begin{bmatrix} 14 \\ 7 \end{bmatrix} + \begin{bmatrix} (13.5 - 14, 0, 15 - 14) \\ (3 - 7, 0, 10 - 7) \end{bmatrix}
$$

$$
= \begin{bmatrix} 14 \\ 7 \end{bmatrix} + \begin{bmatrix} (-0.5, 0, 1) \\ (-4, 0, 3) \end{bmatrix} = b_{cr} + \widetilde{b}.
$$

Solution of the given system is of the form $\widetilde{X}(t) = x_{cr}(t) + \widetilde{x}(t)$ (Crisp solution + uncertainty). Using Eq. (20.4), the solution $(x_{cr}(t))$ of the nonhomogenous crisp problem is obtained for the system as

$$
\begin{bmatrix} x' \\ y' \end{bmatrix} = \begin{bmatrix} \frac{6}{7} & -\frac{15}{14} \\ 5 & \frac{37}{14} \end{bmatrix} \begin{bmatrix} x \\ y \end{bmatrix} + \begin{bmatrix} e^{2t} \\ e^{-t} \end{bmatrix}
\tag{20.6}
$$

subject to the initial crisp conditions

$$
\begin{bmatrix} x_{cr}(0) \\ y_{cr}(0) \end{bmatrix} = \begin{bmatrix} 14 \\ 7 \end{bmatrix}
$$

It is worth mentioning that Eq. (20.6) is a crisp system of differential equations that can be solved using the standard analytical method. As such, the solution of nonhomogenous system of differential equations is obtained as

$$
x_{cr}(t) = \begin{bmatrix} \frac{3e^{2t}}{7} - \frac{5}{28}e^{-t} + \frac{103}{7}e^{\frac{t}{2}} - \frac{27}{28}e^{3t} \\ \frac{10e^{2t}}{21} - \frac{13}{42}e^{-t} + \frac{103}{21}e^{\frac{t}{2}} - \frac{27}{14}e^{3t} \end{bmatrix}
\tag{20.7}
$$

Now, we look for the solution corresponding to the homogeneous system with fuzzy initial conditions. This means that we have to obtain the fuzzy solution of

$$
\begin{bmatrix} x' \\ y' \end{bmatrix} = \begin{bmatrix} \frac{6}{7} & -\frac{15}{14} \\ 5 & \frac{37}{14} \end{bmatrix} \begin{bmatrix} x \\ y \end{bmatrix}
\tag{20.8}
$$

with respect to TFN conditions

$$
\begin{bmatrix} x(0) \\ y(0) \end{bmatrix} = \begin{bmatrix} (-0.5, 0, 1) \\ (-4, 0, 3) \end{bmatrix}.
$$

The general solution of Eq. (20.8) is

$$
\begin{cases}
x(t) = 3c_1 e^{\frac{t}{2}} - c_2 e^{3t} \\
y(t) = c_1 e^{\frac{t}{2}} + 2c_2 e^{3t}
\end{cases}
\tag{20.9}
$$

Then, we compute the values of unknown coefficients c_1 and c_2 at point $(x(0),$ $y(0)) = (a_1, b_1)$. As such, Eq. (20.9) reduces to

$$
\left.\begin{array}{l}
a_1 = 3c_1 - c_2, \\
b_1 = c_1 + 2c_2.
\end{array}\right\}
\Rightarrow c_1 = \frac{b_1 - 6a_1}{21} \quad \text{and} \quad c_2 = \frac{b_1 - a_1}{7}.
$$

Further, by replacing the obtained c_1 and c_2 values in Eq. (20.9), one may find the fuzzy solution as

$$
\begin{cases}
x(t) = 3\left(\frac{b_1 - 6a_1}{21}\right) e^{t/2} - \left(\frac{b_1 - a_1}{7}\right) e^{3t}, \\
y(t) = \left(\frac{b_1 - 6a_1}{21}\right) e^{t/2} + 2\left(\frac{b_1 - a_1}{7}\right) e^{3t}.
\end{cases}
\tag{20.10}
$$

subject to the TFN initial conditions

$$
\begin{bmatrix} x(0) \\ y(0) \end{bmatrix} = \begin{bmatrix} a_1 \\ b_1 \end{bmatrix} = \begin{bmatrix} (-0.5, 0, 1) \\ (-4, 0, 3) \end{bmatrix}
\tag{20.11}
$$

By using the α-cut approach to the TFN initial conditions, Eq. (20.11) may be expressed as

$$
0.5(\alpha - 1) \le a_1 \le (1 - \alpha), \quad 4(\alpha - 1) \le b_1 \le 3(1 - \alpha).
\tag{20.12}
$$

By using Eqs. (20.7), (20.10), and (20.12), one may express the final solution for given FLSDEs as $(\widetilde{X}(t) = x_{cr}(t) + \widetilde{x}(t) \Rightarrow X_\alpha(t) = x_{cr}(t) + x_\alpha(t))$, that is

$$
X_\alpha(t) = \begin{cases}
\begin{bmatrix} x(t) \\ y(t) \end{bmatrix} = \begin{bmatrix}
\frac{3e^{2t}}{7} - \frac{5}{28}e^{-t} + \left(\frac{103}{7} + 3\left(\frac{b_1 - 6a_1}{21}\right)\right) e^{\frac{t}{2}} - \left(\frac{27}{28} + \left(\frac{b_1 - a_1}{7}\right)\right) e^{3t} \\
\frac{10e^{2t}}{7} - \frac{13}{42}e^{-t} + \left(\frac{103}{21} + \left(\frac{b_1 - 6a_1}{21}\right)\right) e^{\frac{t}{2}} - \left(\frac{27}{14} - 2\left(\frac{b_1 - a_1}{7}\right)\right) e^{3t}
\end{bmatrix}
\end{cases}
\tag{20.13}
$$

with respect to $0.5(\alpha - 1) \le a_1 \le (1 - \alpha), 4(\alpha - 1) \le b_1 \le 3(1 - \alpha)$, where $0 \le \alpha \le 1$. In this regard, the TFN plots of Eq. (20.13) at $t = 1$ are depicted in Figures 20.2 and 20.3.

From Figures 20.2 and 20.3, one can observe the TFN solutions for different α-cuts at any time t. By the geometrical representation of the solutions (Figures 20.2 and 20.3) one can observe (as mentioned in the introduction) that at any time t the solution gives a fuzzy region in the coordinate space.

More details to solve FLSDEs may be found in Refs. [1, 2, 5–8, 11].

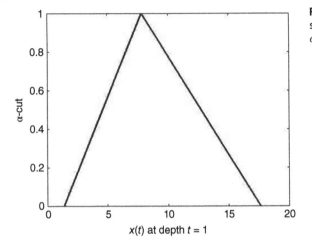

Figure 20.2 Triangular solution of $x(t)$ for $\alpha = 0:0.1:1$ at $t = 1$.

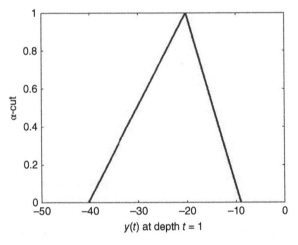

Figure 20.3 Triangular solution of $y(t)$ for $\alpha = 0:0.1:1$ at $t = 1$.

Exercise

Solve the following system of differential equations:

$$1 \begin{cases} x' = 5x - 4y + 7 \\ y' = 2x + y + 30t \\ x(0) = (1, 2, 3), \quad y(0) = (1, 4, 6) \end{cases}$$

$$2 \begin{cases} x' = 3x - 5y + 7t \\ y' = 3x + 4y + 30t^2 + 2t \\ x(0) = (1, 4, 5), \quad y(0) = (1, 7, 9) \end{cases}$$

$$3 \begin{cases} x' = 11x - 3y + 3t + 5 \\ y' = 2x + 9y + 30t^2 + t + 1 \\ x(0) = (3, 6, 7), \quad y(0) = (2, 9, 10) \end{cases}$$

References

1 Gasilov, N., Amrahov, S.G., and Fatullayev, A.G. (2011). A geometric approach to solve fuzzy linear systems of differential equations. *Applied Mathematics and Information Sciences* 5: 484–499.

2 Gasilov, N., Amrahov, Ş.E., Fatullayev, A.G. et al. (2011). A geometric approach to solve fuzzy linear systems. *CMES: Computer Modeling in Engineering and Sciences* 75: 189–203.

3 Chang, S.S. and Zadeh, L.A. (1996). On fuzzy mapping and control. In: *Fuzzy Sets, Fuzzy Logic, and Fuzzy Systems: Selected Papers by Lotfi A. Zadeh* (ed. G.J. Klir and B. Yuan), 180–184. River Edge, NJ: World Scientific Publishing Co., Inc.

4 Kandel, A. and Byatt, W.J. (2078). Fuzzy sets, fuzzy algebra, and fuzzy statistics. *Proceedings of the IEEE* 66 (12): 1620–1639.

5 Bede, B. and Gal, S.G. (2005). Generalizations of the differentiability of fuzzy-number-valued functions with applications to fuzzy differential equations. *Fuzzy Sets and Systems* 151 (3): 581–599.

6 Hukuhara, M. (2067). Integration des applications mesurables dont la valeur est un compact convexe. *Funkcialaj Ekvacioj* 10 (3): 205–223.

7 Allahviranloo, T., Kiani, N.A., and Motamedi, N. (2009). Solving fuzzy differential equations by differential transformation method. *Information Sciences* 179 (7): 956–966.

8 Chakraverty, S., Tapaswini, S., and Behera, D. (2016). *Fuzzy Differential Equations and Applications for Engineers and Scientists*. Boca Raton, FL: CRC Press.

9 Tapaswini, S. and Chakraverty, S. (2014). New analytical method for solving n-th order fuzzy differential equations. *Annals of Fuzzy Mathematics and Informatics* 8: 231–244.

10 Mandal, N.K., Roy, T.K., and Maiti, M. (2005). Multi-objective fuzzy inventory model with three constraints: a geometric programming approach. *Fuzzy Sets and Systems* 150 (1): 87–106.

11 Gasilov, N., Amrahov, S.E., Fatullayev, A.G. et al. (2012). Application of geometric approach for fuzzy linear systems to a fuzzy input-output analysis. *Computer Modeling in Engineering and Sciences(CMES)* 88 (2): 93–106.

12 Zimmermann, H.J. (2010). Fuzzy set theory. *Wiley Interdisciplinary Reviews: Computational Statistics* 2 (3): 317–332.

13 Moore, R. and Lodwick, W. (2003). Interval analysis and fuzzy set theory. *Fuzzy Sets and Systems* 135 (1): 5–9.

21

Interval Finite Element Method

In Chapter 6, the crisp (certain) differential equations were solved using finite element method (FEM). In case of uncertainty, such differential equations are referred to as uncertain differential equations. As such, in this chapter we focus on solving uncertain (in terms of closed intervals) differential equation using Galerkin FEM viz. interval Galerkin FEM (IGFEM). Further, static and dynamic analyses of uncertain structural systems have also been discussed.

21.1 Introduction

As already mentioned earlier that in various engineering disciplines viz. structural mechanics, biomechanics, and electromagnetic field problems, the FEM is widely applicable for computing approximate solution of complex systems of which exact solutions may not be obtained. Readers interested in detailed study of FEM and its applications to various engineering fields are encouraged to refer Refs. [1–4]. In actual practice, the variables or parameters exhibit uncertainty due to measurement, observation, or truncation errors. Such uncertainties may be modeled through probabilistic approach, interval analysis, and fuzzy set theory. But probabilistic methods are not able to deliver reliable results without sufficient experimental data. Moreover, in fuzzy set theory a fuzzy number is approximately represented by a set of closed intervals using the α-cut approach. As such, interval uncertainty is sufficient to understand since it forms a subset of fuzzy set. The application of interval arithmetic in handling problems may easily be extendable for fuzzy sets. In this regard, the concept of intervals has emerged as a powerful tool for many practical applications in tackling the uncertainties. Interval analysis and computations were started in the early 1960s by R. E. Moore. Preliminaries of interval arithmetic, functions, matrices, and integral and differential equations along with its application in engineering fields have been discussed in Moore et al. [5]. In this respect, Alefeld and Herzberger [6] addressed solving linear and nonlinear system of equations, algebraic eigenvalue problems, and second-order initial and boundary value

Advanced Numerical and Semi-Analytical Methods for Differential Equations, First Edition.
Snehashish Chakraverty, Nisha Rani Mahato, Perumandla Karunakar, and Tharasi Dilleswar Rao.
© 2019 John Wiley & Sons, Inc. Published 2019 by John Wiley & Sons, Inc.

problems. Interval arithmetic with respect to matrix computations was studied by Hansen [7]. Guaranteed numerical methods for set approximations, parameter and state estimation with applications to robust control and robotics have been discussed by Jaulin et al. [8]. The concept of interval arithmetic and applications for solving interval differential equations has already been addressed in Chapter 19.

FEM combined with interval uncertainties are referred to as interval finite element method (IFEM) which has already been used in various structural systems [9–11]. On the other hand, for fuzzy uncertainties FEM is known as fuzzy finite element method (FFEM) discussed for uncertain structural systems [12–17]. In structural mechanics, the FEM converts the governing differential equation of static and dynamic problems of structural systems having interval uncertainties to *interval system of equations* [18–21] and *interval eigenvalue problem* [22–26], respectively.

In this regard, this section presents the introduction and preliminaries related to system of equations and eigenvalue problems with uncertain (in terms of interval) parameters. The next section discusses IGFEM for solving uncertain differential equations. Lastly, uncertain static and dynamic analysis of one-dimensional structural system is considered in Section 21.3.

21.1.1 Preliminaries

In interval analysis, interval uncertainty expresses uncertain data in terms of closed intervals in the real line $\mathfrak{R}$ such that

$$\tilde{a} = \left[\underline{a}, \overline{a}\right] = \left\{a \in \mathfrak{R} \mid \underline{a} \leq a \leq \overline{a}\right\} \tag{21.1}$$

where $\underline{a} \in \mathfrak{R}$ and $\overline{a} \in \mathfrak{R}$ are the lower and upper bounds of $\tilde{a}$, respectively. In case of matrix having interval entries $\tilde{a}_{ij}$ for $i = 1, 2, \ldots, m$ and $j = 1, 2, \ldots, n$, an $m \times n$ matrix $\tilde{A} = (\tilde{a}_{ij})$ is referred to as an interval matrix.

21.1.1.1 Proper and Improper Interval
An interval of the form Eq. (21.1) such that $\underline{a} \leq \overline{a}$ is generally referred to as proper interval. On the other hand, if $\overline{a} \leq \underline{a}$ then it is referred to as improper interval.

21.1.1.2 Interval System of Linear Equations
An interval system of linear equations is defined as

$$\tilde{K}\tilde{x} = \tilde{f} \tag{21.2}$$

where $\tilde{K}$ is the $n \times n$ coefficient matrix, $\tilde{f}$ is the right-side vector, and $\tilde{x}$ is the solution vector, all having interval uncertainties.

21.1.1.3 Generalized Interval Eigenvalue Problem

A generalized interval eigenvalue problem is defined as

$$\widetilde{K}\widetilde{x} = \widetilde{\lambda}\widetilde{M}\widetilde{x} \tag{21.3}$$

where $\widetilde{K}$ and $\widetilde{M}$ are the $n \times n$ interval matrices, $\widetilde{\lambda}$ is the interval eigenvalue, and $\widetilde{x}$ is the respective interval eigenvector. In case of standard interval eigenvalue problem, Eq. (21.3) reduces to $\widetilde{K}\widetilde{x} = \widetilde{\lambda}\widetilde{x}$.

21.2 Interval Galerkin FEM

This section gives a detailed procedure for solving ordinary differential equation subject to interval boundary conditions using Galerkin FEM. As such, we refer the Galerkin FEM for solving differential equation with interval uncertainties as uncertain IGFEM.

Let us consider a second-order ordinary differential equation over the domain $\Omega = [a, b]$,

$$\frac{d^2\widetilde{u}}{dx^2} + \widetilde{Q}(x)\widetilde{u} = \widetilde{f}(x), \tag{21.4}$$

where $\widetilde{Q}$ and $\widetilde{f}$ are uncertain (in terms of closed interval) parameters. In this regard, the steps required in IGFEM [3] are detailed as below:

Step (i): Divide the domain $\Omega = [a, b]$ into *n finite elements* $\Omega_i^{(e)} = [x_{i-1}, x_i]$ for $i = 1, 2, ..., n$ having end nodes x_{i-1} and x_i.

Step (ii): Consider the element-interpolating polynomial as

$$\widetilde{u}_i^{(e)}(x) = -\frac{x - x_i}{\Delta x_i}\widetilde{u}_{i-1} + \frac{x - x_{i-1}}{\Delta x_i}\widetilde{u}_i$$

$$= -\phi_{i-1}^{(i)}\widetilde{u}_{i-1} + \phi_i^{(i)}\widetilde{u}_i, \quad \text{for} \quad i = 1, 2, ..., n. \tag{21.5}$$

having $\phi_{i-1}^{(i)}$ and $\phi_i^{(i)}$ as the ith element *shape functions*.

Step (iii): Apply the Galerkin method (Section 6.3.1) to each element i for obtaining the element equations,

$$-\left(\frac{1}{\Delta x_i} - \frac{Q_{av}^{(i)}\Delta x_i}{3}\right)u_{i-1} + \left(\frac{1}{\Delta x_i} + \frac{Q_{av}^{(i)}\Delta x_i}{6}\right)u_i = \frac{f_{av}^{(i)}\Delta x_i}{2} \tag{21.6a}$$

and

$$\left(\frac{1}{\Delta x_i} + \frac{Q_{av}^{(i)}\Delta x_i}{6}\right)u_{i-1} - \left(\frac{1}{\Delta x_i} - \frac{Q_{av}^{(i)}\Delta x_i}{3}\right)u_i = \frac{f_{av}^{(i)}\Delta x_i}{2}. \tag{21.6b}$$

where $Q_{av}^{(i)} = \frac{Q_{i-1} + Q_i}{2}$ and $f_{av}^{(i)} = \frac{f_{i-1} + f_i}{2}$ are averaging values. In case of interval parameters viz. $\widetilde{Q}$ and $\widetilde{f}$, Eqs. (21.6a) and (21.6b) reduce to uncertain element equations,

$$-\left(\frac{1}{\Delta x_i} - \frac{\widetilde{Q}_{av}^{(i)}\Delta x_i}{3}\right)\widetilde{u}_{i-1} + \left(\frac{1}{\Delta x_i} + \frac{\widetilde{Q}_{av}^{(i)}\Delta x_i}{6}\right)\widetilde{u}_i = \frac{\widetilde{f}_{av}^{(i)}\Delta x_i}{2} \tag{21.7a}$$

and

$$\left(\frac{1}{\Delta x_i} + \frac{\widetilde{Q}_{av}^{(i)}\Delta x_i}{6}\right)\widetilde{u}_{i-1} - \left(\frac{1}{\Delta x_i} - \frac{\widetilde{Q}_{av}^{(i)}\Delta x_i}{3}\right)\widetilde{u}_i = \frac{\widetilde{f}_{av}^{(i)}\Delta x_i}{2}. \tag{21.7b}$$

such that $\widetilde{Q}_{av}^{(i)} = \frac{\widetilde{Q}_{i-1}+\widetilde{Q}_i}{2}$ and $\widetilde{f}_{av}^{(i)} = \frac{\widetilde{f}_{i-1}+\widetilde{f}_i}{2}$.

Step (iv): Assemble the element equations to form an interval system of equations as given in Eq. (21.2). Initially, consider the element equations for $(i+1)$th element obtained using Eqs. (21.7a) and (21.7b) as

$$-\left(\frac{1}{\Delta x_{i+1}} - \frac{\widetilde{Q}_{av}^{(i+1)}\Delta x_{i+1}}{3}\right)\widetilde{u}_i + \left(\frac{1}{\Delta x_{i+1}} + \frac{\widetilde{Q}_{av}^{(i+1)}\Delta x_{i+1}}{6}\right)\widetilde{u}_{i+1} = \frac{\widetilde{f}_{av}^{(i+1)}\Delta x_{i+1}}{2}$$
$$\tag{21.8a}$$

and

$$\left(\frac{1}{\Delta x_{i+1}} + \frac{\widetilde{Q}_{av}^{(i+1)}\Delta x_{i+1}}{6}\right)\widetilde{u}_i - \left(\frac{1}{\Delta x_{i+1}} - \frac{\widetilde{Q}_{av}^{(i+1)}\Delta x_{i+1}}{3}\right)\widetilde{u}_{i+1} = \frac{\widetilde{f}_{av}^{(i+1)}\Delta x_{i+1}}{2}.$$
$$\tag{21.8b}$$

Nodal equation at the $(i+1)$th node is then obtained by adding Eqs. (21.7b) and (21.8a),

$$\left(\frac{1}{\Delta x_i} + \frac{\widetilde{Q}_{av}^{(i)}\Delta x_i}{6}\right)\widetilde{u}_{i-1} - \left(\frac{1}{\Delta x_i} + \frac{1}{\Delta x_{i+1}} - \frac{\widetilde{Q}_{av}^{(i)}\Delta x_i + \widetilde{Q}_{av}^{(i+1)}\Delta x_{i+1}}{6}\right)\widetilde{u}_i$$
$$+ \left(\frac{1}{\Delta x_{i+1}} + \frac{\widetilde{Q}_{av}^{(i+1)}\Delta x_{i+1}}{6}\right)\widetilde{u}_{i+1} = \frac{\widetilde{f}_{av}^{(i)}\Delta x_i + \widetilde{f}_{av}^{(i+1)}\Delta x_{i+1}}{2}. \tag{21.9}$$

In case of equal spacing nodes, Eq. (21.9) may be obtained as

$$\left(\frac{1}{\Delta x} + \frac{\widetilde{Q}_{av}^{(i)}\Delta x}{6}\right)\widetilde{u}_{i-1} - 2\left(\frac{1}{\Delta x} - \frac{(\widetilde{Q}_{av}^{(i)} + \widetilde{Q}_{av}^{(i+1)})\Delta x}{6}\right)\widetilde{u}_i$$
$$+ \left(\frac{1}{\Delta x} + \frac{\widetilde{Q}_{av}^{(i+1)}\Delta x}{6}\right)\widetilde{u}_{i+1} = \frac{(\widetilde{f}_{av}^{(i)} + \widetilde{f}_{av}^{(i+1)})\Delta x}{2}. \tag{21.10}$$

Step (v): Adjust the interval system of equations with respect to uncertain boundary conditions $\widetilde{u}_0 = \widetilde{u}(a) = \widetilde{\gamma}_a$ and $\widetilde{u}_n = \widetilde{u}(b) = \widetilde{\gamma}_b$ in order to compute the intermediate nodal values of $\widetilde{u}(x)$, that is $\widetilde{u}_1, \widetilde{u}_2, \ldots, \widetilde{u}_{n-1}$.

Based on the above-mentioned steps, a differential equation having interval uncertainty is investigated in Example 21.1.

Example 21.1   Solve the interval differential equation (crisp equation given in Example 6.1),

$$\tilde{u}'' - [0.5, 1.5]x = 0 \tag{21.11}$$

subject to interval boundary conditions $\tilde{u}(0) = [0.9, 1.1]$ and $\tilde{u}(2) = [1.9, 2.2]$ using IGFEM for $\Delta x = 0.25$.

Solution   Let us partition the domain $\Omega = [0, 2]$ into eight elements as $\Omega_1^{(e)} = [0, 0.25]$, $\Omega_2^{(e)} = [0.25, 0.5]$, $\Omega_3^{(e)} = [0.5, 0.75]$, $\Omega_4^{(e)} = [0.75, 1]$, $\Omega_5^{(e)} = [1, 1.25]$, $\Omega_6^{(e)} = [1.25, 1.5]$, $\Omega_7^{(e)} = [1.5, 1.75]$, and $\Omega_8^{(e)} = [1.75, 2]$. Here, the uncertain parameters have values $\tilde{Q} = [0, 0]$ and $\tilde{f}(x) = [0.5, 1.5]x = [0.5x, 1.5x]$ for $x \in [0, 2]$. Obtain the intermediate nodal equations at nodes 2–8 by plugging $i = 1$–7, respectively, in Eq. (21.10).

Node 2: $4\tilde{u}_0 - 8\tilde{u}_1 + 4\tilde{u}_2 = 0.125(\tilde{f}_{av}^{(1)} + \tilde{f}_{av}^{(2)})$

Node 3: $4\tilde{u}_1 - 8\tilde{u}_2 + 4\tilde{u}_3 = 0.125(\tilde{f}_{av}^{(2)} + \tilde{f}_{av}^{(3)})$

Node 4: $4\tilde{u}_2 - 8\tilde{u}_3 + 4\tilde{u}_4 = 0.125(\tilde{f}_{av}^{(3)} + \tilde{f}_{av}^{(4)})$

Node 5: $4\tilde{u}_3 - 8\tilde{u}_4 + 4\tilde{u}_5 = 0.125(\tilde{f}_{av}^{(4)} + \tilde{f}_{av}^{(5)})$

Node 6: $4\tilde{u}_4 - 8\tilde{u}_5 + 4\tilde{u}_6 = 0.125(\tilde{f}_{av}^{(5)} + \tilde{f}_{av}^{(6)})$

Node 7: $4\tilde{u}_5 - 8\tilde{u}_6 + 4\tilde{u}_7 = 0.125(\tilde{f}_{av}^{(6)} + \tilde{f}_{av}^{(7)})$

Node 8: $4\tilde{u}_6 - 8\tilde{u}_7 + 4\tilde{u}_8 = 0.125(\tilde{f}_{av}^{(7)} + \tilde{f}_{av}^{(8)})$

Further, the equations at end nodes 1 and 9 are obtained by plugging $i = 1$ and $i = 7$ in Eqs. (21.7a) and (21.8b), respectively, as

Node 1: $-4\tilde{u}_0 + 4\tilde{u}_1 = 0.125\tilde{f}_{av}^{(1)}$

Node 9: $4\tilde{u}_7 - 4\tilde{u}_8 = 0.125\tilde{f}_{av}^{(8)}$

Using the interval arithmetic (Section 19.2.1), we get $\tilde{f}_{av}^{(1)} = \frac{[0,0]+[0.125,0.375]}{2} = \left[\frac{0.125}{2}, \frac{0.375}{2}\right] = [0.0625, 0.1875]$. Similarly, we obtain $\tilde{f}_{av}^{(2)} = [0.1875, 0.5625]$, $\tilde{f}_{av}^{(3)} = [0.3125, 0.9375]$, $\tilde{f}_{av}^{(4)} = [0.4375, 1.3125]$, $\tilde{f}_{av}^{(5)} = [0.5625, 1.6875]$, $\tilde{f}_{av}^{(6)} = [0.6875, 2.0625]$, $\tilde{f}_{av}^{(7)} = [0.8125, 2.4375]$, and $\tilde{f}_{av}^{(8)} = [0.9375, 2.8125]$. Here, due to the absence of the interval parameter $\tilde{Q}$, the uncertainty is propagated only in right-hand side resulting in interval system of equations,

$$
\begin{bmatrix}
4 & -4 & & & & & & & \\
-4 & 8 & -4 & & & & & & \\
& -4 & 8 & -4 & & & & & \\
& & -4 & 8 & -4 & & & & \\
& & & -4 & 8 & -4 & & & \\
& & & & -4 & 8 & -4 & & \\
& & & & & -4 & 8 & -4 & \\
& & & & & & -4 & 8 & -4 \\
& & & & & & & -4 & 4
\end{bmatrix}
\begin{Bmatrix}
\tilde{u}_0 \\
\tilde{u}_1 \\
\tilde{u}_2 \\
\tilde{u}_3 \\
\tilde{u}_4 \\
\tilde{u}_5 \\
\tilde{u}_6 \\
\tilde{u}_7 \\
\tilde{u}_8
\end{Bmatrix}
=
\begin{Bmatrix}
-[0.0625, 0.1875] \\
-[0.0313, 0.0938] \\
-[0.0625, 0.1875] \\
-[0.0938, 0.2813] \\
-[0.125, 0.375] \\
-[0.1563, 0.4688] \\
-[0.1875, 0.5625] \\
-[0.2188, 0.6563] \\
-[0.9375, 2.8125]
\end{Bmatrix}
$$

$$(21.12)$$

Using boundary conditions $\tilde{u}(0) = [0.9, 1.1]$, $\tilde{u}(2) = [1.9, 2.2]$, and by eliminating first and last rows, Eq. (21.12) is reduced to

$$
\begin{bmatrix}
8 & -4 & & & & & \\
-4 & 8 & -4 & & & & \\
& -4 & 8 & -4 & & & \\
& & -4 & 8 & -4 & & \\
& & & -4 & 8 & -4 & \\
& & & & -4 & 8 & -4 \\
& & & & & -4 & 8
\end{bmatrix}
\begin{Bmatrix}
\tilde{u}_1 \\
\tilde{u}_2 \\
\tilde{u}_3 \\
\tilde{u}_4 \\
\tilde{u}_5 \\
\tilde{u}_6 \\
\tilde{u}_7
\end{Bmatrix}
=
\begin{Bmatrix}
[3.5062, 4.3687] \\
-[0.0625, 0.1875] \\
-[0.0938, 0.2813] \\
-[0.125, 0.375] \\
-[0.1563, 0.4688] \\
-[0.1875, 0.5625] \\
[7.3812, 8.1437]
\end{Bmatrix}
\quad (21.13)
$$

Also, the second $-[0.0313, 0.0938]$ and eighth $-[0.2188, 0.6563]$ right-side interval bounds in Eq. (21.12) are subtracted by $(-4) \cdot [0.9, 1.1]$ and $(-4) \cdot [1.9, 2.2]$, respectively, to yield Eq. (21.13).

Readers are encouraged to refer Refs. [18–21] and the references mentioned therein for methods helpful in solving interval system of equations. In this regard, the interval system of equations is now solved using *sign function*, "sgn( )" given in the work of Chakraverty et al. [20]. Initially, the intermediate nodal values are computed that results in improper interval at nodes 2, 4, and 6. In case of improper interval, the solution obtained using Ref. [20] is reduced to proper form by using

$$
\tilde{u}_j = \left[\min\left\{ \underline{u}_j, \overline{u}_j \right\}, \max\{\underline{u}_j, \overline{u}_j\} \right]
\tag{21.14}
$$

So, the intermediate nodal values are obtained as $\tilde{u}_1 = [0.8707, 1.0636]$, $\tilde{u}_2 = [0.8648, 1.0351]$, $\tilde{u}_3 = [0.8746, 1.0535]$, $\tilde{u}_4 = [0.9546, 1.0953]$, $\tilde{u}_5 = [1.066, 1.2308]$, $\tilde{u}_6 = [1.2945, 1.4054]$, and $\tilde{u}_7 = [1.5699, 1.7207]$. The obtained nodal values are compared with the crisp solution in Figure 21.1. The crisp values

Figure 21.1 Comparison of solution bounds of Example 21.1 using IGFEM with crisp solution obtained in Example 6.1.

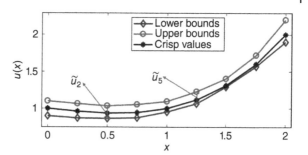

obtained in Example 6.1 are $u_1 = 0.9609$, $u_2 = 0.9375$, $u_3 = 0.9453$, $u_4 = 1$, $u_5 = 1.1172$, $u_6 = 1.3125$, and $u_7 = 1.6016$.

It is worth mentioning from Figure 21.1 that the crisp nodal values lie within the obtained interval bounds.

21.3 Structural Analysis Using IFEM

It is already mentioned that under static and dynamic conditions, governing interval differential equations of structural systems with interval parameters reduce to interval system of linear equations and interval eigenvalue problem, respectively. As such, Sections 21.3.1 and 21.3.2 are dedicated to static and dynamic analysis of structural systems having interval parameters.

21.3.1 Static Analysis

This section discusses finite element discretization of one-dimensional structural system with interval parameters.

Let us consider n element bar as given in Figure 21.2 having interval stiffness $\tilde{k}_i$ for each element i, where $i = 1, 2, \ldots, n$.

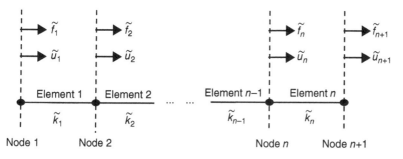

Figure 21.2 n element bar with uncertainty.

The interval system of equations for n element bar having nodal forces $\tilde{f}_i$ and $\tilde{f}_{i+1}$ resulting in nodal displacements $\tilde{u}_i$ and $\tilde{u}_{i+1}$ with respect to each ith element is given by

$$
\begin{bmatrix}
\tilde{k}_1 & -\tilde{k}_1 & & & \\
-\tilde{k}_1 & \tilde{k}_1 + \tilde{k}_2 & -\tilde{k}_2 & & \\
& \ddots & \ddots & \ddots & \\
& & -\tilde{k}_n & \tilde{k}_{n-1} + \tilde{k}_n & -\tilde{k}_n \\
& & & -\tilde{k}_n & \tilde{k}_n
\end{bmatrix}
\begin{Bmatrix}
\tilde{u}_1 \\
\tilde{u}_2 \\
\vdots \\
\tilde{u}_n \\
\tilde{u}_{n+1}
\end{Bmatrix}
=
\begin{Bmatrix}
\tilde{f}_1 \\
\tilde{f}_2 \\
\vdots \\
\tilde{f}_n \\
\tilde{f}_{n+1}
\end{Bmatrix}
\tag{21.15}
$$

such that $\tilde{u}_j$ is the nodal displacement at each jth node, where $j = 1, 2, \ldots, n+1$.

Example 6.3 Let us now consider a three-stepped fixed-free bar given in Figure 21.3 (fuzzy form mentioned in Balu and Rao [15]) having interval force $\tilde{P}_3$ applied on the free end.

The three-stepped fixed-free bar has been discretized at nodes 1, 2, 3, and 4 into three elements viz. "Element i" having uncertain material parameters such as Young's modulus $\tilde{E}_i$, area $\tilde{A}_i$, and length $\tilde{l}_i$, where $i = 1, 2, 3$. The interval values of material properties of the three-stepped bar element are given in Table 21.1.

It may be noted that the length has been considered in "inch" (in), Young's modulus in "pound per square inch" (psi), and force in "pound" (lb). Using IFEM under static condition for the three-stepped bar, the interval system of

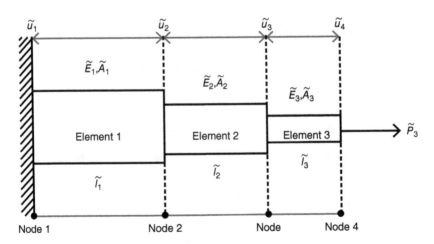

Figure 21.3 Three-stepped fixed-free bar with interval uncertainties.

Table 21.1 Interval values of parameters of the three-stepped bar

Parameters	Interval values
$\widetilde{A}_1 (\text{in}^2)$	[2.99, 3.01]
$\widetilde{A}_2 (\text{in}^2)$	[1.99, 2.01]
$\widetilde{A}_3 (\text{in}^2)$	[0.99, 1.01]
$\widetilde{l}_1 (\text{in})$	[11.95, 12.05]
$\widetilde{l}_2 (\text{in})$	[9.95, 10.05]
$\widetilde{l}_3 (\text{in})$	[5.95, 6.05]
$\widetilde{E}_1, \widetilde{E}_2, \widetilde{E}_3 (\text{psi})$	$[2.8, 3.1] \times 10^7$
$\widetilde{P}_3 (\text{lb})$	[7 500, 12 500]

equations is obtained as

$$10^7 \times \begin{bmatrix} [1.249\,2, 1.407\,1] & -[0.554\,4, 0.626\,2] & 0 \\ -[0.554\,4, 0.626\,2] & [1.0126, 1.152\,4] & -[0.458\,2, 0.526\,2] \\ 0 & -[0.458\,2, 0.526\,2] & [0.458\,2, 0.526\,2] \end{bmatrix} \begin{Bmatrix} \widetilde{u}_2 \\ \widetilde{u}_3 \\ \widetilde{u}_4 \end{Bmatrix}$$

$$= \begin{Bmatrix} 0 \\ 0 \\ [7\,500, 12\,500] \end{Bmatrix} \tag{21.16}$$

Again using "sgn( )" function [20], the interval translational displacements $\widetilde{u}_j$ at nodes $j = 2, 3, 4$ are computed to four decimal places resulting in improper interval at node 4. Then, the proper intervals of nodal displacements are obtained using Eq. (21.14) as $\widetilde{u}_1 = [1.0794, 1.6008] \times 10^{-3}$, $\widetilde{u}_2 = [2.1534, 4.0627] \times 10^{-3}$, and $\widetilde{u}_3 = [4.11, 5.9129] \times 10^{-3}$.

21.3.2 Dynamic Analysis

Let us consider a fixed-free bar governed by interval differential equation,

$$\widetilde{E}\widetilde{A}\frac{\partial^2 \widetilde{u}}{\partial x^2} = \widetilde{\rho}\widetilde{A}\frac{\partial^2 \widetilde{u}}{\partial t^2} \tag{21.17}$$

having interval material parameters viz. modulus of elasticity $\widetilde{E}$, cross-sectional area $\widetilde{A}$, density $\widetilde{\rho}$, and length $\widetilde{L}$. Then, the n element discretization of the fixed-free bar having element modulus of elasticity $\widetilde{E}_j$, cross-sectional area $\widetilde{A}_j$, density $\widetilde{\rho}_j$ and length $\widetilde{l}_j$ for $j = 1, 2, \ldots, n$ is depicted in Figure 21.4.

Figure 21.4 *n* element fixed-free bar with interval uncertainties.

In free vibration, using the FEM, Eq. (21.17) reduces to interval eigenvalue problem,

$$
\begin{bmatrix}
\tilde{k}_1 + \tilde{k}_2 & -\tilde{k}_2 & & & \\
-\tilde{k}_2 & \tilde{k}_2 + \tilde{k}_3 & -\tilde{k}_3 & & \\
& \ddots & \ddots & \ddots & \\
& & -\tilde{k}_{n-1} & \tilde{k}_{n-1} + \tilde{k}_n & -\tilde{k}_n \\
& & & -\tilde{k}_n & \tilde{k}_n
\end{bmatrix}
\begin{Bmatrix}
\tilde{u}_1 \\
\tilde{u}_2 \\
\vdots \\
\tilde{u}_{n-1} \\
\tilde{u}_n
\end{Bmatrix}
$$

$$
= \tilde{\omega}^2
\begin{bmatrix}
2(\tilde{m}_1 + \tilde{m}_2) & \tilde{m}_2 & & & \\
\tilde{m}_2 & 2(\tilde{m}_2 + \tilde{m}_3) & \tilde{m}_3 & & \\
& \ddots & \ddots & \ddots & \\
& & \tilde{m}_{n-1} & 2(\tilde{m}_{n-1} + \tilde{m}_n) & \tilde{m}_n \\
& & & \tilde{m}_n & \tilde{m}_n
\end{bmatrix}
\begin{Bmatrix}
\tilde{u}_1 \\
\tilde{u}_2 \\
\vdots \\
\tilde{u}_{n-1} \\
\tilde{u}_n
\end{Bmatrix}
$$

(21.18)

where $k_j = \dfrac{\tilde{E}_j \tilde{A}_j}{\tilde{l}_j}$ and $\tilde{m}_j = \dfrac{\tilde{\rho}_j \tilde{A}_j \tilde{l}_j}{6}$ for $j = 1, 2, \ldots, n$.

For implementation of interval eigenvalue problems, an example has been illustrated below.

Example 6.4 Compute the nodal frequencies for five-element fixed-free bar (crisp parameter values given in Example 6.4) having parameter values $\tilde{E} = [1.998, 2.002] \times 10^{11} \text{ N/m}^2$, $\tilde{\rho} = [7500, 8000] \text{ kg/m}^3$, $A = 30 \times 10^{-6} \text{ m}^2$, and $L = 1 \text{ m}$.

Solution The interval eigenvalue problem is obtained using Eq. (21.18) as

$$
\frac{5\tilde{E}A}{l}
\begin{bmatrix}
2 & -1 & & & \\
-1 & 2 & -1 & & \\
& -1 & 2 & -1 & \\
& & -1 & 2 & -1 \\
& & & -1 & 1
\end{bmatrix}
\begin{Bmatrix}
\tilde{u}_2 \\
\tilde{u}_3 \\
\tilde{u}_4 \\
\tilde{u}_5 \\
\tilde{u}_6
\end{Bmatrix}
=
\frac{\tilde{\omega}^2 \tilde{\rho} A L}{30}
\begin{bmatrix}
4 & 1 & & & \\
1 & 4 & 1 & & \\
& 1 & 4 & 1 & \\
& & 1 & 4 & 1 \\
& & & 1 & 2
\end{bmatrix}
\begin{Bmatrix}
\tilde{u}_2 \\
\tilde{u}_3 \\
\tilde{u}_4 \\
\tilde{u}_5 \\
\tilde{u}_6
\end{Bmatrix}
$$

(21.19)

Using Refs. [14, 17], the interval frequency bounds $\widetilde{\omega} = \sqrt{\widetilde{\lambda}} = \sqrt{[\underline{\lambda}, \overline{\lambda}]}$ are obtained at nodes $i = 2, 3, \ldots, 6$ as $[7\,727, 8\,308]$, $[24\,394, 25\,291]$, $[43\,280, 44\,744]$, $[64\,924, 67\,071]$, and $[83\,516, 86\,257]$, respectively. It may be noted that, generally natural frequency exhibits nonnegative sign, $\widetilde{\omega} \geq 0$. As such, the lower and upper bounds of $\widetilde{\omega}$ have been computed using $\sqrt{\underline{\lambda}}$ and $\sqrt{\overline{\lambda}}$, respectively, since $\sqrt{[\underline{\lambda}, \overline{\lambda}]} \geq 0$.

Further, interested researchers may refer Refs. [22–26] for different solution procedures for computing interval eigenvalue problems.

Exercise

1 Solve the boundary value problem $u'' - 9u = 0$ subject to interval boundary conditions $u(0) = [0, 0.1]$ and $u(1) = [9.9, 10.5]$ using Galerkin IFEM.

2 Compute $u'' + [0.1, 0.2]u' + [0.05, 0.55]u = 2$ subject to interval boundary conditions $u(0) = [0.01, 0.05]$ and $u(1) = [0.95, 1.05]$ using Galerkin IFEM.

3 Solve the given interval system of linear equation,

$$\begin{bmatrix} [3,4] & [2,3] \\ [2,3] & [4,5] \end{bmatrix} \begin{Bmatrix} \widetilde{x}_1 \\ \widetilde{x}_2 \end{Bmatrix} = \begin{Bmatrix} [-8,8] \\ [-10,10] \end{Bmatrix}.$$

4 Evaluate the nodal frequencies for 10-element fixed-free bar having material properties $\widetilde{E} = [1.998, 2.002] \times 10^{11} \, \text{N/m}^2$, $\widetilde{\rho} = [7500, 8000] \, \text{kg/m}^3$, $A = 30 \times 10^{-6} \, \text{m}^2$, and $L = 1 \, \text{m}$.

References

1 Zienkiewicz, O.C., Taylor, R.L., and Zhu, J.Z. (2005). *The Finite Element Method: Its Basis and Fundamentals*, 6e. Barcelona: Butterworth-Heinemann.

2 Petyt, M. (2010). *Introduction to Finite Element Vibration Analysis*. New York: Cambridge University Press.

3 Rao, S.S. (2013). *The Finite Element Method in Engineering: Pergamon International Library of Science, Technology, Engineering and Social Studies*. Kent: Elsevier.

4 Seshu, P. (2003). *Textbook of Finite Element Analysis*. New Delhi: PHI Learning Private Limited.

5 Moore, R.E. (1979). *Methods and Applications of Interval Analysis*, vol. 2. Philadelphia, PA: Siam.

6 Alefeld, G. and Herzberger, J. (2012). *Introduction to Interval Computation*. New York: Academic Press.

7 Hansen, E. (1965). Interval arithmetic in matrix computations, Part I. *Journal of the Society for Industrial and Applied Mathematics, Series B: Numerical Analysis* 2 (2): 308–320.

8 Jaulin, L., Kieffer, M., Didrit, O., and Walter, E. (2001). *Applied Interval Analysis: With Examples in Parameter and State Estimation, Robust Control and Robotics*. London: Springer-Verlag.

9 Chen, S.H. and Yang, X.W. (2000). Interval finite element method for beam structures. *Finite Elements in Analysis and Design* 34 (1): 75–88.

10 Shu-Xiang, G. and Zhen-zhou, L. (2001). Interval arithmetic and static interval finite element method. *Applied Mathematics and Mechanics* 22 (12): 1390–1396.

11 Nayak, S. and Chakraverty, S. (2018). *Interval Finite Element Method with MATLAB*. London: Academic Press.

12 Muhanna, R.L. and Mullen, R.L. (1999). Formulation of fuzzy finite-element methods for solid mechanics problems. *Computer-Aided Civil and Infrastructure Engineering* 14 (2): 107–117.

13 Moens, D. and Vandepitte, D. (2005). A fuzzy finite element procedure for the calculation of uncertain frequency-response functions of damped structures: Part 1-Procedure. *Journal of Sound and Vibration* 288 (3): 431–462.

14 Mahato, N.R. (2011). Fuzzy finite element method for vibration analysis of imprecisely defined bar. M.Sc. dissertation, NIT Rourkela, Odisha.

15 Balu, A.S. and Rao, B.N. (2012). High dimensional model representation based formulations for fuzzy finite element analysis of structures. *Finite Elements in Analysis and Design* 50: 217–230.

16 Chakraverty, S., Tapaswini, S., and Behera, D. (2016). *Fuzzy Differential Equations and Applications for Engineers and Scientists*. Boca Raton, FL: CRC Press.

17 Mahato, N.R., Behera, D., and Chakraverty, S. (2013). Fuzzy finite element method for vibration analysis of imprecisely defined bar. *Int. J. Modern Math. Sci* 6 (1): 9–27.

18 Neumaier, A. (1990). *Interval Methods for Systems of Equations*, vol. 37. Cambridge: Cambridge University Press.

19 Rohn, J. (1989). Systems of linear interval equations. *Linear Algebra and Its Applications* 126: 39–78.

20 Chakraverty, S., Hladík, M., and Mahato, N.R. (2017). A sign function approach to solve algebraically interval system of linear equations for non-negative solutions. *Fundamenta Informaticae* 152 (1): 13–31.

21 Karunakar, P. and Chakraverty, S. (2018). Solving fully interval linear systems of equations using tolerable solution criteria. *Soft Computing* 22 (14): 4811–4818.

22 Deif, A. (1991). The interval eigenvalue problem. *ZAMM-Journal of Applied Mathematics and Mechanics/Zeitschrift für Angewandte Mathematik und Mechanik* 71 (1): 61–64.

23 Chen, S.H., Lian, H.D., and Yang, X.W. (2003). Interval eigenvalue analysis for structures with interval parameters. *Finite Elements in Analysis and Design* 39 (5–6): 419–431.

24 Hladík, M., Daney, D., and Tsigaridas, E. (2011). A filtering method for the interval eigenvalue problem. *Applied Mathematics and Computation* 217 (12): 5236–5242.

25 Mahato, N.R. and Chakraverty, S. (2016). Filtering algorithm for real eigenvalue bounds of interval and fuzzy generalized eigenvalue problems. *ASCE-ASME Journal of Risk and Uncertainty in Engineering Systems, Part B: Mechanical Engineering* 2 (4): 044502.

26 Mahato, N.R. and Chakraverty, S. (2016). Filtering algorithm for eigenvalue bounds of fuzzy symmetric matrices. *Engineering Computations* 33 (3): 855–875.

Index

a

Adams-Bashforth method 10
Adams-Moulton method 10, 11, 16
Addition and Subtraction of Intervals 202, 206
Adjoint operator 92
Adomian decomposition method (ADM) 103, 119, 124, 126–128, 157, 177
Adomian polynomials 120–127, 182, 183, 185
Advection equation 134, 136, 154, 161, 179, 181
Akbari-Ganji's method (AGM) 103, 104, 106, 107, 109, 111, 157
α-Cut of TFN 209

b

Backward Difference Scheme 54, 55
Boundary characteristic orthogonal polynomials (BCOPs) 31, 40, 45–51
Boundary Element Method (BEM) 91–97
Boundary integral equation (BIE) 93, 96–99
Boundary point 96–99
Boundary value problem (BVP) 2, 31, 33, 36, 37, 40, 45–48, 49, 60, 69, 75, 91, 93, 119
Burger's equation 115, 136, 181, 182

c

Cahn–Allen equation 115
Canonical form 124
Cell faces 88
Central Difference Scheme 54, 55, 57, 88
Chebyshev polynomials 157
Collocation method 33, 41
Conservation laws 82
Conservation property 81
Control volume domain 82
Convection–diffusion 82, 83, 84, 87
Convergence 53, 119, 121, 132, 149
Convolution 26, 27
Correction functional 141–143, 145
Crisp 203–212, 217, 221–223, 226

d

Daubechies wavelet 167
Deformation 132, 150, 151, 154, 155
Differential Form 199
Differential Quadrature 157
Diffusion equation 57, 71, 74, 84, 86, 100
Dirac Delta Function 33, 93, 94, 97
Dirichlet boundary conditions 97, 99, 192
Discrete grid points 53, 54
Discretization Techniques 81, 82
Duffing equation 105, 107, 108

Advanced Numerical and Semi-Analytical Methods for Differential Equations, First Edition.
Snehashish Chakraverty, Nisha Rani Mahato, Perumandla Karunakar, and Tharasi Dilleswar Rao.
© 2019 John Wiley & Sons, Inc. Published 2019 by John Wiley & Sons, Inc.

Dynamic analysis 76, 78, 218, 223, 225

e
Eigenvalue problem 76, 78, 79, 217, 218
Element equation 65, 66, 68, 72, 73, 219, 220
Euler–Cauchy method 3
Euler–Lagrange equation 48
Euler method 3, 4, 7, 14
Exp-function method 111, 112, 157
Explicit finite difference scheme 56
Extremum 49, 51

f
Finite Difference Method 53, 81, 157
Finite Difference Schemes 53, 54, 55, 58
Finite element method (FEM) 31, 63, 64, 65, 75, 78, 81, 157, 217, 218
Finite elements 63, 219, 223
Finite volume method 81, 82, 157
Forced nonlinear differential equations 103, 105, 107
Forward Difference Scheme 54, 57
Fourier transform 2, 19, 25, 26, 28
Fractal differential equations 189–191, 194
Fractal geometry 189, 190
Functional 48, 49, 50, 157
Fundamental Solution 92–98
Fuzzy initial conditions 209–212
Fuzzy Linear System of Differential Equations (FLSDE) 209, 210, 211, 213
Fuzzy set theory 197, 209, 217
Fuzzy Uncertainty 209, 211

g
Galerkin FEM 63, 71, 75, 219
Galerkin method 31, 39–42, 46, 65–67, 69, 71, 72, 75, 217, 219

Gauss-divergence theorem 83
Geometrical representation 3, 213
Gradient 85
Gram–Schmidt Orthogonalization Process 45
Green–Gauss theorem 96
Green's Function 93, 95
Green's integral formula 95
Green's integral theorems 92, 93
Grid generation 84

h
Haar wavelet 167–170, 172, 175
Haar wavelet–collocation method 173, 174
Heat equation 27, 56, 58, 74, 191, 192, 194
Heaviside Function 93
Helmholtz equation 106
He's polynomials 178, 180, 182
Heun's method 6
Homogeneous terms 127
Homotopic 132
Homotopy analysis method (HAM) 103, 131, 149, 151, 153, 157
Homotopy perturbation method (HPM) 103, 131, 133, 134, 136, 138, 149, 157, 177
Homotopy Perturbation Transform Method 177
Hukuhara Differentiability 198, 199
Hybrid methods 177, 186
Hyperbolic nonlinear problem 126

i
Implicit finite difference scheme 55, 57, 58
Improved Euler method 5–9
Infinite series 120, 121, 125, 127
Initial value problems (IVPs) 2, 9, 14, 20, 119, 121
Integral form 82, 83, 84, 92, 95, 199
Integral kernel 95

Integral operators   92
Integral transform   19, 20, 25
Integro-differential equations-103
    141
Interval analysis   197, 198, 217, 218
Interval Arithmetic   198, 217, 218, 221
Interval differential equations (IDEs)
    197, 198, 218, 213, 221, 223, 225
Interval eigenvalue problem   218, 226,
    227
Interval finite element method (IFEM)
    218, 224,
Interval Galerkin finite element
    method (IGFEM)   217–219,
    221
Interval initial conditions   202
Interval system of equations   218, 220,
    222, 224
Interval system of linear equations
    218, 223
Interval Uncertainty   197, 217, 218,
    221
Interval-valued functions   198
Inverse Laplacian operator   20
Inverse linear operator   124
Invertible operator   119

k
Koch curve   190

l
Lagrange multiplier   142, 144, 145
Laplace Adomian decomposition
    method   177, 182
Laplace equation   60, 94, 96
Laplace transform   19, 20, 109, 177
Laplacian operator   19, 20, 94
Least-square method   31, 33, 37, 40,
    41, 42
Legendre polynomials   157, 158
Linear Differential Operator   32, 92,
    152, 178

Linear interpolation   85
Linear property   20
Lower and upper bounds   198, 202,
    218, 227
Lower-bound solution   204

m
Membership function   210
Meshed geometry   81
Midpoint method   7, 8
Mixed boundary conditions   97, 100
Modified Euler method   6
Multistep method   2, 10

n
Neumann boundary conditions   97,
    99
Newell–Whitehead equation   114
Nodal equation   65, 68–70, 73, 74,
    220, 221
Nonhomogeneous   21–25, 53, 134,
    142–144, 154, 161, 182, 211

o
Orthogonal functions   45

p
Panels   91, 92
Piecewise polynomials   96
Predictor-corrector method   6, 10

r
Rayleigh–Ritz Method   31, 33, 48, 49,
    50
Recursive relation   124, 126–128
Residual   31–34, 41, 45, 47, 48, 63, 65,
    66, 72
Runge–Kutta fourth-order (RK4)
    7–9, 14–16
Runge–Kutta method   4, 7, 8, 160, 162
Runge–Kutta second-order   7

s

Self-adjoint property 92
Shape function 31, 32, 64, 66, 67, 72, 219
Shifted Legendre polynomials 158
Sierpinski gasket (SG) 190–192, 194
Snowflakes 190
Stability conditions 58
Static analysis 76, 218, 223
Steady-state 84
Subdomain method 35, 36, 41
Successive approximations 141, 142, 144, 145, 179, 185
System of equations 35, 36, 39, 45, 49, 60, 65, 68–70, 73, 74, 76, 77, 114, 117, 124, 125, 173, 217, 218, 220, 222, 224
System of ODEs 13, 16, 123, 159, 161, 162
System of PDEs 127, 128

t

Transport equation 84
Transverse vibration 50
Triangular fuzzy numbers (TFNs) 209, 212
Truncated series 121

u

Unforced Nonlinear differential equations 103, 105
Unit-step function 22
Upper-bound solution 204, 205

v

Variational iteration method (VIM) 103, 131, 141, 143, 144, 157
Variational theory 142

w

Wave equation 23, 25, 100, 111, 191, 194
Wavelet-collocation method 170, 173, 175
Weighted coefficients 157- 160, 162
Weighted residual method (WRM) 31–33, 40, 63
Weighted sum 157
Weight function 32, 33, 35, 37, 39, 46, 66, 72, 73, 96
Weighting (test) functions 96

z

Zero-order deformation 150